Application of Knowledge Management to Utilities

The mission of the Awwa Research Foundation (AwwaRF) is to advance the science of water to improve the quality of life. Funded primarily through annual subscription payments from over 1,000 utilities, consulting firms, and manufacturers in North America and abroad, AwwaRF sponsors research on all aspects of drinking water, including supply and resources, treatment, monitoring and analysis, distribution, management, and health effects.

From its headquarters in Denver, Colorado, the AwwaRF staff directs and supports the efforts of over 500 volunteers, who are the heart of the research program. These volunteers, serving on various boards and committees, use their expertise to select and monitor research studies to benefit the entire drinking water community.

Research findings are disseminated through a number of technology transfer activities, including research reports, conferences, videotape summaries, and periodicals.

Application of Knowledge Management to Utilities

Prepared by:
Jeffrey S. Rosen, **David B. Miller**, **Kristyn B. Stevens**, **Ayhan Ergul**, and **Jose A.H. Sobrinho**
Technology Planning and Management Corporation
Mill Wharf Plaza, Suite 208, Scituate, MA 02066

and

Michelle M. Frey
McGuire Environmental Consultants
1620 Market Street, Suite 3E, Denver, CO 80202

and

Larry Pinkstaff
Curry Controls Company
1019 Pipkin Road, Lakeland, FL 33811

Sponsored by:
Awwa Research Foundation
6666 West Quincy Avenue
Denver, CO 80235-3098

Published by the

Awwa Research Foundation
Advancing the Science of Water®

IWA Publishing

DISCLAIMER

This study was funded by the Awwa Research Foundation (AwwaRF). AwwaRF assumes no responsibility for the content of the research study reported in this publication or for the opinions or statements of fact expressed in the report. The mention of trade names for commercial products does not represent or imply the approval or endorsement of AwwaRF. This report is presented solely for informational purposes.

Published by IWA Publishing, Alliance House, 12 Caxton Street, London SW1H 0QS, UK

Telephone: +44 (0) 20 7654 5500; Fax: +44 (0) 20 7654 5555; Email: publications@iwap.co.uk;
Web: **www.iwapublishing.com**

AwwaRF report number 90895F. Originally published by AwwaRF for its subscribers in 2003.
IWA Publishing version published 2004.

ISBN 1 84339 851 6

CONTENTS

TABLES

FIGURES

FOREWORD

The Awwa Research Foundation (AwwaRF) is a nonprofit corporation dedicated to the implementation of a research effort to help utilities respond to regulatory requirements and traditional high-priority concerns of the industry. The research agenda is developed through a process of consultation with subscribers and drinking water professionals. Under the umbrella of a Strategic Research Plan, the Research Advisory Council prioritizes the suggested projects based upon current and future needs, applicability, and past work, and forwards the recommendations to the Board of Trustees for final selection. The foundation also sponsors research projects through the unsolicited proposal process; the Collaborative Research, Research Applications, and Tailored Collaboration programs; and various joint research efforts with organizations such as the United States Environmental Protection Agency (U.S. EPA), the U.S. Bureau of Reclamation, and the Association of California Water Agencies.

This publication is a result of one of these sponsored studies, and it is hoped that its findings will be applied in communities throughout the world. The following report serves not only as a means of communicating the results of the water industry's centralized research program but also as a tool to enlist the further support of the nonmember utilities and individuals.

The Foundation's staff and large cadre of volunteers, who willingly contribute their time and expertise, manage projects closely from their inception to the final report. The foundation serves a planning and management function and awards contracts to other institutions such as water utilities, universities, and engineering firms. The funding for this research effort comes primarily from the Subscription Program, through which water utilities subscribe to the research program and make an annual payment proportionate to the volume of water they deliver. Consultants and manufacturers subscribe based on their annual billings. The program offers a cost-effective and fair method for funding research in the public interest.

The Foundation's research agenda addresses a broad spectrum of water supply issues: resources, treatment and operations, distribution and storage, water quality and analysis, toxicology, economics, and management. The ultimate purpose of the coordinated effort is to assist water suppliers to provide the highest possible quality of water economically and reliably. The true benefits are realized when the results are implemented at the utility level. The Foundation's trustees are pleased to offer this publication as a contribution toward that end.

This project focuses on how new information tools, collectively referred to as Knowledge Management tools, can be best employed by drinking water utilities to help organize, access and apply water quality data to utility operations. This document summarizes the current state of Knowledge Management. This project demonstrates the methods by which data that are collected by a variety of sources can be processed, aggregated and applied to support a wide array of decision support activities in an efficient manner. The goal of this report is to provide a process for improvement in water quality data management and daily operations of drinking water utilities.

Edmund G. Archuleta, P.E.
Chair, Board of Trustees
Awwa Research Foundation

James F. Manwaring, P.E.
Executive Director
Awwa Research Foundation

ACKNOWLEDGMENTS

The authors of this report are indebted to the following water utilities and individuals for their cooperation and participation in this project:

City of Golden, CO
Newport News Waterworks, VA
Massachusetts Water Resources Authority, MA
East Bay Municipal Utility District, CA
City of Tucson, AZ

In addition, the authors greatly appreciate the assistance and advice of the AwwaRF Project Manager, Frank Blaha, Denver, Colorado and the Project Advisory Committee (PAC)—including Richard Kirby, Anglian Water, Peterborough, Cambridgeshire, United Kingdom; Ronald Hunsinger, East Bay Municipal Utility District, Oakland, California; Stephen Stanley, EPCOR Water Services, Edmonton, Alberta, Canada; and Richard Donner, Seattle Public Utilities, Seattle, Washington. This project benefited from the series of PAC reviews including workshop participation and critique of all documents. We are grateful to the PAC for their continued commitment and involvement in this project, as their assistance was invaluable.

The authors also would like to acknowledge Stephen Estes-Smargiassi, Massachusetts Water Resources Authority, Boston, Massachusetts for his participation in the PAC workshop. Pamela Curren supported the review, production and coordination of this document.

ACKNOWLEDGEMENTS

The authors of this report are [illegible]

[illegible]

EXECUTIVE SUMMARY

Historically, the main focus of water supply reliability has been assuring the quantity of available water supplies (Cromwell et al. 1997). The majority of the information collected was of limited use and application. In recent years, assurance of the quality of available water supplies has been added as a goal (Cromwell et al. 1997). With new requirements and a perception that the data collected is a resource that can improve efficiency and water quality, data access has taken on a new urgency. In order to meet these increased standards of efficiency and customer service, water suppliers are finding it necessary to redefine their business approaches. Management, analysis, and dissemination of data and information are becoming strategic approaches to improve efficiencies and support daily decisions by operators and managers. However, the management and dissemination of the data are major challenges.

Knowledge Management and Business Intelligence are processes and tools that allow businesses to increase workflow efficiency while reducing employee workload and labor hours. These processes and tools help employees manage, access, and use information as a resource in and their application to the drinking water industry. A literature review of Knowledge Management tools included case studies and five on-site utility evaluations. In addition, a pilot system-was developed to show how a utility could make water quality knowledge more available to senior managers in a succinct, useful manner.

What is Knowledge Management?

There is a broad range of perspectives on the definition and scope of Knowledge Management. American Productivity and Quality Center (1996) and Firestone (1998b) view Knowledge Management as the strategies and processes that are capable of identifying, capturing, and maintaining an organization's knowledge. The SAS Institute (1999e) defines Knowledge Management as a system-based approach capable of capturing and leveraging institutional memory. Butler (1999) defines Knowledge Management as an integrated approach comprised of processes, technologies, attitudes, and systems that promotes and provides for adaptation and survival of the organization in a rapidly changing environmental workplace. Despite the various definitions, the main goal of Knowledge Management is to identify, capture, manage, analyze, and extract knowledge from various sources within an organization and place it into a structured, organized format that can be shared. While this organized format may need to be refined over time, Knowledge Management requires the use of underlying data, standards, and methodology.

It is important to distinguish between Knowledge Management and Information Management. While the definition of Knowledge Management is not clear, a distinction between Knowledge Management and Information Management is quite clear. Information Management is the organization of data and information into a logical structure that facilitates the capture, quality assurance, management, analysis, and dissemination of the data and information. It is the development of a structure and rules for organizing the data and information. Knowledge Management, on the other hand, is the development of relationships between the different data structures. It includes integration of many databases into a higher-level understanding of how the data and information relate to one another. While Information Management involves the organization and application of data and information to decision making, Knowledge Management relates the results of the analyses to one another, developing insights and wisdom about the relationships of the results. Knowledge Management captures the insights of individuals and automates these insights into the processing of information and into observing and extracting relationships between different data in underlying information systems.

A Knowledge Management System (KMS) is not a package that can be purchased like a statistical or accounting package. A KMS is something developed and applied as a solution for a particular organization. It consists of many components selected for the particular capabilities that can be integrated to meet specific requirements. It is a solution based on many tools. KMSs often interact with databases, e-mail, document management systems, and operating systems directories. The support technologies that assist with Knowledge Management include data warehousing, records and document management, data mining, On Line Analytical Processing (OLAP), portals, enterprise resource planning, and customer relationship management.

It is important to note that Knowledge Management improvements are not only contingent on the installation of new hardware and software, but also depend on a change in culture and use of data and information. The installation of these tools in the absence of a change in the conduct of business will not have a significant effect, and in some cases may result in poorer performance. Knowledge Management requires a significant change in a utilities organizational culture to be successful. This cultural change is often more difficult than the investment and the challenge of gathering and organizing all of the required information. A commitment must be made to become an information-based organization. System implementation needs to include extensive training in the specific application of the system especially for middle and upper managers.

RESEARCH OBJECTIVES

The Awwa Research Foundation (AwwaRF) funded this study to help improve the data management and operations of drinking water utilities. This report presents an evaluation of the applicability of emerging Knowledge Management tools for improving the capture, management, analysis, and communication of data and information related to water quality. The results are based on the integration of a literature review, intensive case studies, and application of the tools in a pilot project. The four objectives of the study included:

1. Review the state of Knowledge Management as an industry.
2. Illustrate how Knowledge Management can be used in a drinking water utility.
3. Review how Knowledge Management is being applied in utilities.
4. Develop a pilot (Knowledge Management implementation) for a drinking water utility.

APPROACH

Review the State of Knowledge Management as an Industry

This project included a literature review on Knowledge Management and Business Intelligence and associated applications. See Appendix A for a comprehensive description of the Knowledge Management literature review.

Illustrate How Knowledge Management Can Be Used in a Drinking Water Utility

We evaluated case studies to gain insight into how Knowledge Management is applied in a utility. We also conducted requirements analyses at drinking water utilities to evaluate current data management needs and identify requirements that could be effectively addressed by the application of a variety of Knowledge Management tools. The Utilities reviewed included the City of Golden, Colorado Water Department; City of Tucson, Arizona Water Department; Newport News Waterworks, Virginia; Massachusetts Water Resources Authority, Massachusetts; and East Bay Municipal Utility District, California. Work groups conducted at each participating utility covered the following information topics:

- Information developed by each group within a utility
- Information obtained from other groups
- How information was communicated between groups
- Reports and products produced by each group
- Barriers to information needs

Review How Knowledge Management is Being Applied in Utilities

Each evaluation lent further insight into how Knowledge Management is being applied in utilities. Based on all of this information, the project team developed this report to provide a process for improvement in water quality data management and daily operations of drinking water utilities. This report will discuss and demonstrate how Knowledge Management can be applied to facilitate these improvements.

Develop a Pilot (Knowledge Management Implementation) for a Drinking Water Utility

To demonstrate the potential utility of Knowledge Management tools, we developed a pilot software system. The case studies have shown that the current drinking water utilities are somewhat limited in their ability to utilize and share the large amount of data at their disposal. Most of the case study utilities were at the D1 (Raw Data) level (see Figure 3.4). The pilot was an attempt to move utilities to the next level and demonstrate the project recommendations. In order to limit the pilot development, we placed the focus on the tools that can utilize the data currently available at Tucson Water Department and an audience geared to senior management. The community supported by the pilot probably will be limited, but the application of the approach is intended for any user within a utility, given the customization ability built into the pilot system.

CONCLUSIONS

Most utilities exhibited a range of strategies for implementing information systems. In general, information systems were both centralized and distributed. However, many of the systems were implemented to meet general needs and often did not meet specific needs of the utility. The general information requirements were consistent among all of the utilities.

Several utilities already had sophisticated information systems that worked well for a specific application. However, these systems often failed when they were integrated into a utility-wide information system. The reasons for these failures were varied, but included a lack of robust requirements analysis prior to implementation. Other major failures resulted from attempts to solve all information processing needs with a single system. Laboratory Information Management Systems (LIMS) and maintenance management systems often failed. These systems often were perfect for their intended roles but could not meet all of the water quality information processing needs effectively.

This project demonstrated how a requirements analysis can be done at a utility with limited time and resources. It also identified and categorized the required building blocks that must be considered by drinking water utilities if they are to improve their information use and upgrade operations, compliance reporting, and decision support. If the functionality of the pilot is appropriate and meets the needs of utilities, this project has developed a model that can be applied by other utilities.

FUTURE RESEARCH

Future research needs include the development of industry-wide standards and integrated water quality databases. Access to water quality information is often scattered across an organization without an easy method of combination. Research needs also include Web-based interfaces to provide better user access to the integrated information systems. Finally, improved processing, integration and dissemination of SCADA data will provide the real-time access to data often only available to operations staff.

CHAPTER 1
INTRODUCTION

BACKGROUND

Drinking water utilities are complex business organizations that generate a significant amount of varying information, including business, operations, water quality, planning, and communications with stakeholders. Recently, national drinking water regulations have required more data collection and reporting (U.S. EPA 1998). The amount of data collected, along with the added requirements of federal and state regulations, has resulted in an increased interest in data utilization in the management of drinking water utilities.

The drinking water industry has long been rich in raw data but poor in the aggregated information derived from these data. Routinely, utilities perform numerous measurements of water resources management and quality of source water, finished water, and water delivered to customers. Utilities measure a wide range of data through a variety of monitoring strategies. Field technicians and operators perform in-situ measurements, while automated measurements collected via SCADA systems are made using on-line monitors. Grab samples are taken from a variety of places and are analyzed in utility and commercial laboratories. Municipal, state and federal agencies collect and disseminate hydrologic, weather, demographic, and emergency data critical to effective management of drinking water utilities. Many of the data described above are used daily to make operational decisions regarding water resource management, water treatment, and distribution strategies. Failure of accurate and timely reporting of some of these data can result in compliance violations that can be costly and embarrassing for utilities. If managed, aggregated, and stored effectively, these data could help determine management strategies for repeated situations.

In spite of the importance of data generated, collected and available to drinking water utilities, few manage the data effectively and consistently across the utility. In many cases, data are generated, used immediately, then archived in formats that are inaccessible to all but the most technical users within a utility. Data management within a single utility can range from storage in relational databases to data sheets kept on clipboards in a treatment plant or booster station.

Analytical results within a single laboratory often are stored in completely unrelated formats and systems.

Pressures from new regulations and privatization are forcing utilities to re-evaluate their strategies for information processing and reporting. Regulators are considering approaches that base drinking water standards on 95th percentiles and moving averages spanning monthly, quarterly, or annual intervals. In order to meet these new regulations and compliance requirements, utilities will need to improve their data management, aggregation and reporting capabilities. Electronic Data Interchange (EDI) initially will present utilities with a unique way to submit compliance reports with minimal errors and delays. Eventually, EDI probably will become a requirement that will have major effects on utilities management and operations due to information processing needs.

These requirements, new to the drinking water industry, have been utilized effectively in the electric and gas industries for many years. Integrated information systems are routinely installed and include operations, maintenance, monitoring, LIMS, GIS, staffing, and historical records of related data. Explosive growth and development in the IT arena present a unique opportunity to the drinking water community as it aggressively enters the information age. The required functionality can be implemented while the broad range of potential solutions could lead to success in implementing systems.

Drinking water utilities share many characteristics of large commercial businesses and large government agencies. They have universal tasks, including billing, human resources, maintenance, customer support, operations, and engineering. Unlike many commercial businesses, they have major compliance requirements that involve careful monitoring and data management. As these requirements continue to expand and as the drinking water business becomes more competitive, proper use and leverage of available information will improve a utility's competitiveness, compliance reporting, and water quality.

In an effort to assist drinking water utilities to understand the status of information technologies that can support management, access, and use of an array of information distributed throughout a utility, we reviewed Knowledge Management and its associated applications. The main goals of Knowledge Management are to identify, capture, manage, analyze, and extract knowledge from various sources within an organization and place it into a structured, organized format that can be shared. While this organized format may need to be refined over time, Knowledge Management requires the use of underlying data, standards, and methodology.

RATIONALE AND OBJECTIVES

We designed project objectives to complete the following:

- Review the state of Knowledge Management as an industry
- Illustrate how Knowledge Management might be used in a drinking water utility
- Review how Knowledge Management is being applied in utilities
- Develop a prototype Knowledge Management implementation for a drinking water utility

We conducted a critical review of Knowledge Management and its associated tools, and reviewed Knowledge Management literature to determine its functionality and applicability to the drinking water industry. We also evaluated Knowledge Management for functionality, case study application, and its applicability to drinking water utilities.

Simultaneously, we conducted an evaluation of information requirements of drinking water utilities, including a high-level requirements analysis, using standard requirements methods, at five utilities. Based on the analyses, we identified integrated water quality requirements for drinking water utilities. We identified barriers, and proposed recommendations and solutions to the barriers. We developed a pilot application of Knowledge Management for a drinking water utility to demonstrate the recommendations and the application of Knowledge Management for drinking water utilities.

REPORT ORGANIZATION

This report contains four sections. Chapter 1 presents an introduction, which details the background, statement of the problem, and the project objectives. Chapter 2 presents Methods and Materials, which detail the approaches taken in this process. Chapter 3 summarizes the results and provides a brief discussion. Chapter 4 presents a Summary and Conclusions, including the relevance of this project to the drinking water community. This report also contains three appendices: Appendix A, Final Report for Phase 1 Knowledge Management and Business Intelligence Review, Appendix B, the blank questionnaire used in the utility evaluations, and Appendix C, the Utility Analysis Report.

CHAPTER 2
METHODS AND MATERIALS

We conducted this project in three distinct phases as indicated in Table 2.1.

Table 2.1

Project Tasks

Project Phase	Task
Task 1	Assess Knowledge Management Techniques: 1.1 Develop an annotated bibliography of references 1.2 Document application of Knowledge Management in utilities
Task 2	Develop a consolidated reference on water quality management
Task 3	Assess the application of Knowledge Management techniques to water quality management

TASK 1. ASSESS KNOWLEDGE MANAGEMENT TECHNIQUES

The objective of this task was to conduct a literature search that reviewed and detailed the current state of Knowledge Management and Business Intelligence and a summary of how these methodologies are currently being used in the drinking water industry. This literature review synthesized AwwaRF documents, online white papers, gray literature, and published journal articles regarding Knowledge Management and Business Intelligence tools including data warehousing, OLAP, data mining, and data portals. This review also included customer support, record management, and planning.

A matrix of applications supporting these Knowledge Management tools was developed to assist an organization as it begins to develop a Knowledge Management System by selecting functions needed to improve the organization's efficiency. Also, the application of these Knowledge Management tools to utilities, including drinking water, wastewater, and energy, was evaluated. Finally, an annotated bibliography was developed to show the relevance and quality of the references cited. All of these products are included in Appendix A.

TASK 2. DEVELOP A CONSOLIDATED REFERENCE ON WATER QUALITY MANAGEMENT

This task included a synthesis of results from the Knowledge Management literature review. We compiled a report was compiled that evaluated Information Management and knowledge-based technology for the capture, management, and dissemination of data and information for drinking water utilities. We completed a summary describing the capabilities and future paths for Knowledge Management (Appendix A). In addition, we conducted a high-level requirements analysis at the following five drinking water utilities:

- Massachusetts Water Resources Authority, Massachusetts
- City of Tucson, Arizona
- City of Golden, Colorado
- East Bay Municipal Utility District, California
- Newport News Waterworks, Virginia

We conducted intensive high level requirements analyses, including questionnaires, focus group meetings, and detailed report-outs, for each utility. Table 2.2 indicates the dates that each work group was conducted at each facility:

Table 2.2

Site Evaluation Schedule

Partner Utility	Dates of Site Evaluations
Massachusetts Water Resource Authority	December 12–14, 2000
City of Tucson	May 8–13, 1999
City of Golden	July 19–21, 2000
Newport News Waterworks	September 18–20, 2000
East Bay Municipal Utility District	March 26–29, 2001

Table 2.3 presents basic information about the utilities including the populations served, source type, treatment type, Millions of Gallons per day (MGD), average consumption, number of employees, number of IS staff, and the ratio of IT staff to employees.

Table 2.3

Utility Information

Utility	MWRA	City of Golden	City of Tucson	East Bay MUD	Newport News Waterworks
Location	Massachusetts	Colorado	Arizona	California	Virginia
Population Served	2,500,000	16,200	675,000	1,200,000	360,000
Source Type	Surface Water	Surface Water	Ground Water	Surface Water	Surface Water
Treatment Type	Chemical	Conventional	Chemical	Variety	Conventional
MGD	275	7	120	215	51
Average Consumption (gallons per capita per day)	110	370	178	179	142
Number of Employees	1572	35	575	2000	370
Number of IS Staff	90	5	12	42	7
IT/Total Employees	5.7%	14.3%	2.1%	2.1%	1.9%

The evaluations conducted at each utility covered a range of topics including:

- Information developed by each group within a utility
- Information obtained from other groups
- How information is communicated between groups
- Reports and products produced by each group
- Barriers to information needs

Each session began by sending the blank questionnaire (Appendix B) to essential staff members. We scheduled and conducted one-hour interviews with staff by functional group (or division). Once completed, the project team developed a synthesis of information. The project team gave a preliminary report in a PowerPoint presentation at each utility, for review by the utility. When necessary, the team incorporated revisions to observations, opportunities, and recommendations. Finally, we drafted a written case study report for review with customized aggregates for each utility. This focused on information access, Information Management, data quality, data processing, field data, and barriers. Appendix C includes each report.

TASK 3. ASSESS THE APPLICATION OF KNOWLEDGE MANAGEMENT TECHNIQUES TO WATER QUALITY MANAGEMENT

The objective of this task was to summarize how the identified requirements could be addressed using Knowledge Management tools and to develop a pilot system to demonstrate it. The first tasks involved summarizing the current state of the practice and capabilities of cutting-edge Information Technology for capturing, processing, and disseminating data, information, and textual and graphic information. In addition, the project team established general requirements for Information Management and use in drinking water utilities. Ultimately, this type of system should dramatically improve the management and use of information within utilities.

The project team selected Tucson Water as the utility to provide the water quality data for this pilot, as the members were already familiar with the information systems at Tucson from a previous project. This facilitated the development of the pilot because the understanding of the infrastructure, information needs, and personnel had already been established. To expedite the production of the pilot system, the core infrastructure was based on Open Source technology except for the database systems, which used an Oracle database for the warehouse. The key technology was the application server, ZOPE, and a powerful scripting language, Python. ZOPE is an object-oriented system that can be used to rapidly design web-based applications especially when joining to database systems. Using open source systems allowed for cost savings as well as the ability to utilize third party extensions. These extensions gave improved graphing capability, document access for indexing, and the ability to communicate with e-mail systems.

CONCLUSION

The pilot system intentionally has limitations on the breadth of issues it tries to solve, but if a utility were to use the same technology, much of the pilot system could be part of a production application. The migration from the pilot system to a production version will require substantial expansion to fully populate the notions demonstrated in the pilot. The reluctance of many utilities to build production systems on open source technology may result in extra cost and effort to replicate the capabilities of the pilot using commercial software. The extra costs also need to include the third party extensions as they are likely be commercial software as well and not part of a base package.

CHAPTER 3
RESULTS

LITERATURE REVIEW

Commercial businesses already have recognized the need for Knowledge Management (termed Business Intelligence in this domain) to efficiently gain a competitive advantage. Drinking water utilities need to examine ways to better utilize their limited resources the same as "for profit" businesses. In an effort to assist drinking water utilities to understand the status of information technologies that can support management, the project team reviewed the access and use of an array of information distributed throughout a utility, the use of Knowledge Management techniques, and associated applications. The main goal of Knowledge Management is to identify, capture, manage, analyze, and extract knowledge from various sources within an organization and place it into a structured, organized format that can be shared. While this organized format may need to be refined over time, Knowledge Management requires the use of underlying data, standards, and methodology. There is consensus that no single system or tool can provide a KMS and is not software that can be purchased like a statistical package or accounting package. Designing such a system requires careful thought and reorganization of workflow. This is something developed and applied as a solution for a particular organization. Therefore, Knowledge Management should not be viewed as a tool, but rather as a set of tools and capabilities or "Middleware" (Figure 3.1).

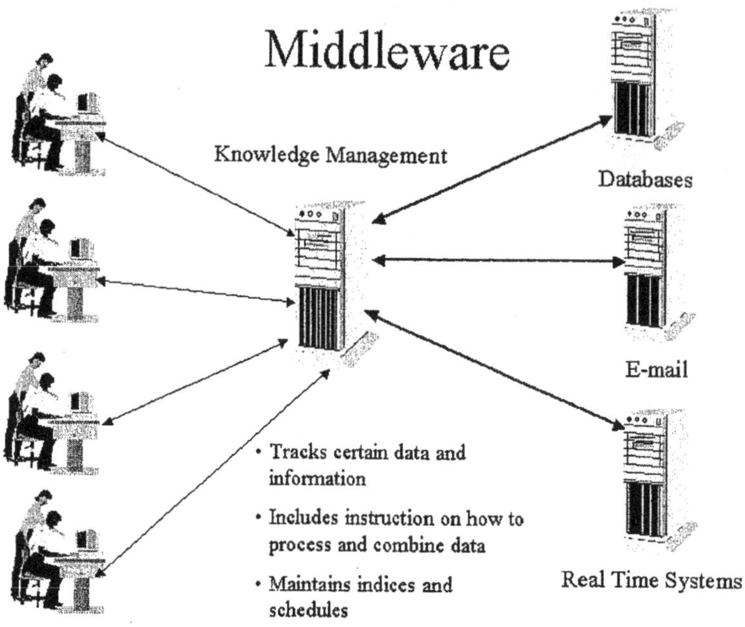

Figure 3.1 Knowledge Management as "Middleware"

A Knowledge Management System consists of many components selected for the particular capabilities that can be integrated to meet particular requirements. Knowledge portals are essential aspects to using this system by providing each viewer his or her own customized access to knowledge (Figure 3.2). Portals can be used effectively to act as the window into a database or data warehouse, as well as consolidating the knowledge needed specifically for each individual.

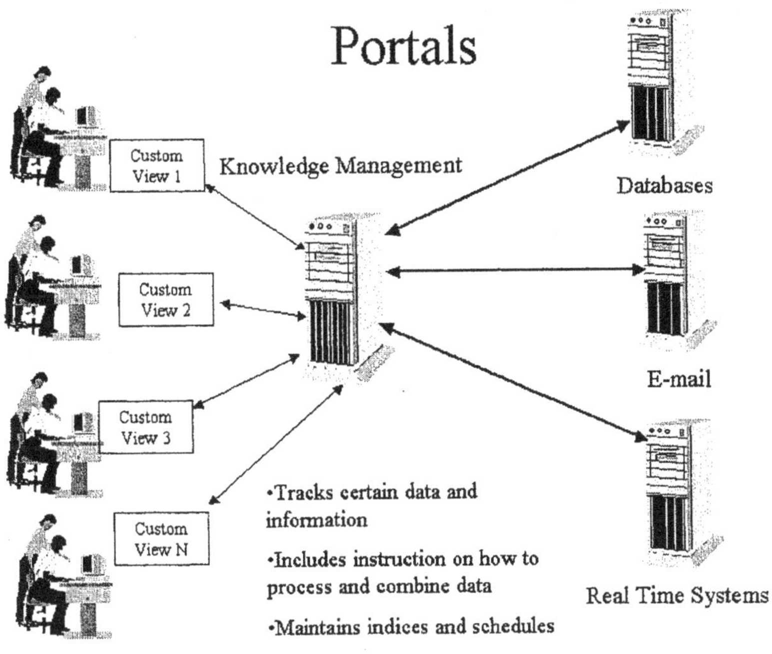

Figure 3.2 The use of Portals in a Knowledge Management System

Besides access to general information, automation of aspects of the system such as reports, triggers, and alarms may be warranted. Computing systems tend to be more exact than people in catching predefined events and are less prone to an error of oversight (Figure 3.3).

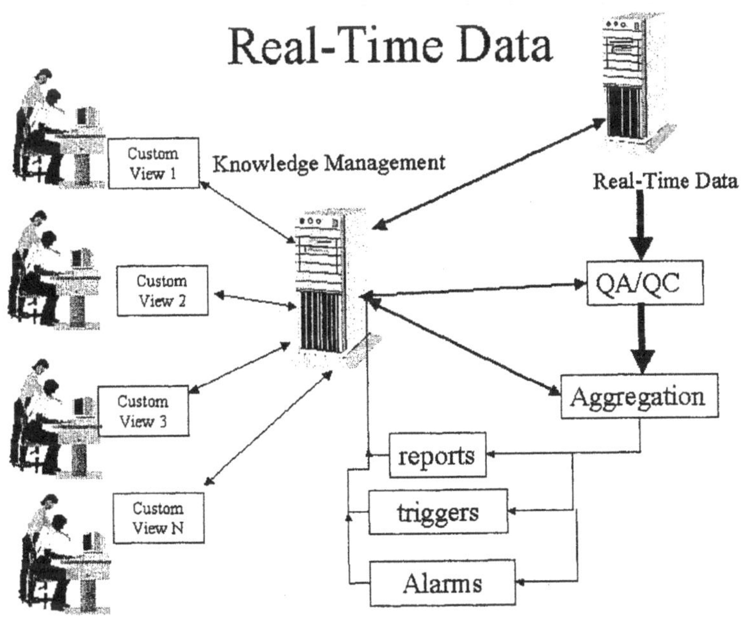

Figure 3.3 Real-Time Data Access Through Knowledge Management

TRANSFORMATION OF DATA INTO KNOWLEDGE

The effective use of data and information requires an understanding of data quality and aggregation methods that have been utilized. Figure 3.4 illustrates the conversion of data into knowledge and its relationship to prospective users from technicians and operators to the public. As data is processed, analyzed, and migrated up the left triangle, decisions are made regarding aggregation strategies, data exclusion due to quality assurance issues, and statistical approaches. Therefore, information systems' content requires increasingly more documentation in order for content to be used effectively. As a general trend, the users become less indoctrinated in the detailed uses of databases and computer systems as the categories go up the right hand triangle, requiring more aggregations, simpler interfaces, and concise summaries of the documentation. The required documentation is implicit in each component of Figure 3.4. The data in each compartment needs to be carefully documented so users understand how the data can and should be used. Documentation that indicates how the information aggregation is completed and detailing the content of the new level must be included. At each transition, as data are aggregated, quality assurance must be performed to ensure that the information has been aggregated and documented properly. Many users assume that quality assurance needs to be done only at the D1 and D2 levels. This is not the case. However, all aggregations involve calculations, assumptions, and summarizations. Once the aggregates are created and stored, it is difficult for most users to go back and check that the aggregation was completed properly. Routine quality assurance and documentation ensures that the aggregation is both understood and completed properly. If quality assurance is not completed, the aggregates are used with little to no question. This can result in serious errors.

The current state of Information Management of most utilities is at the D1 – D2 level. Utilities need to begin exploring the development of aggregate information systems that bring together summarized and simplified information from a variety of the core systems listed above. The aggregate information system can be either a physical or a virtual system that is rigorously maintained and quality-assured. The aggregate information systems should be made available to the majority of the utility personnel. Underlying all these types of systems must be a sound, comprehensive information system with clear documentation concerning quality and a clear understanding of the nature of the data contained in the information system. This system may be one or more data warehouses or possibly direct access to a set of databases. Data warehouses are primarily a performance solution. Additional capabilities draw upon this data using Knowledge Management tools such as OLAP, data mining, records and document management, portals, Enterprise Resource Planning (ERP), and Customer Relationship Management (CRM).

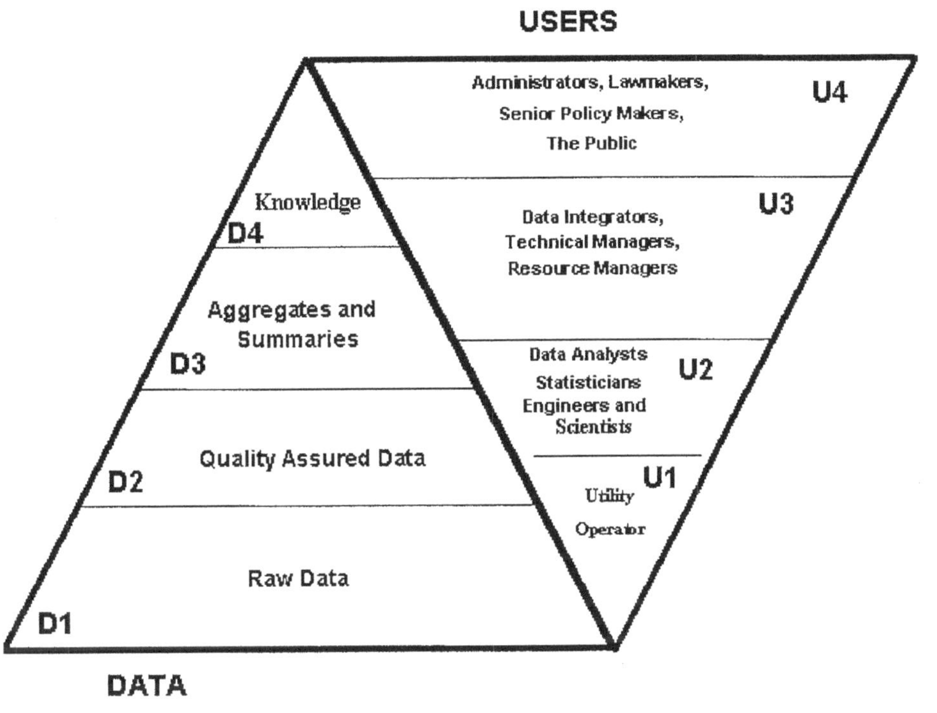

Figure 3.4 Information Triangle

Once the data warehouse is built, OLAP provides a methodology for supporting automatic storage, reformatting, and processing of the data going into the warehouse. OLAP is an analytical information processing technology, generally integrated to the data warehouse, with the ability to examine data from multiple dimensions and is performed "online" and interactively (SAS Institute 1996b). Through the use of transformations and calculations on existing data, OLAP creates business information. This information either can be stored as part of the warehouse or presented to the user.

While data warehouses store and organize data, there is still a need to filter efficiently through the vast warehouses of data. Data mining is the main process by which trends can be identified and evaluated, which turns data into information, and provides visual representations for data exploration, statistical analysis and modeling, and an assessment of the results (SAS Institute 1998a). Data mining could also be used to address Quality Assurance (QA) issues. This should occur at every level (D1 through D4). Currently, utilities quality assure data at the raw data level (D1), but tend to break down at the data aggregation level (D3). Data mining could assist with range checking, outlier analysis, and quality control charting. Portals also would draw on this data to assist in selective views of the information, and records management would assist in finding and tracking information stored "somewhere" in the system.

Records and documents are among the most valuable assets within an organization. Records can be any information essential to a company and may reside in numerous locations, including on-site and off-site locations and document archives. Records need to be stored and retrieved and may have history that needs to be preserved. Records can include (but are not limited to) databases, e-mail, documents, images, graphs, database records, data files, spreadsheets, scanned documents, letters, permits, etc. Anything that can be stored in an electronic file is considered a record. Documents usually are analog records that are stored either as hard copy or as scanned images. These images can be both electronic and analog (e.g., microfiche). Documents can include items like letters, permits, applications, maps, faxes, PDF files, and microfilm. Some documents and records have a useful life span that should not be exceeded.

Records and document management help determine when and how a document should be destroyed. The document management system itself can be either an electronic or a paper system. Document management applications manage (e.g., creation, capture, indexing, and retrieval), store, and distribute documents throughout their lifecycle. They also add version control, shared workflow, and structured communications.

Portals can be used effectively as the window into a warehouse. They also can be used to automate Knowledge Management strategies customized to individual users while accessing a single warehouse. Modern Enterprise Information Portals (EIPs) allow an organization to customize the view of an underlying information system or warehouse (Firestone 1999a). Each group or perspective within a corporation or a wider user community can be customized to deliver the subset of information required in a format and level of content detail appropriate to a particular group or individual. EIPs provide total access (essentially a single gateway) to multiple forms of critical data as a starting point for a user. Some types of portals even support e-mail and scheduling functions. It is important to note that most EIPs require a dedicated server for accessing data (Firestone 1999b). They usually depend on a host of underlying technologies (e.g., databases, transactional systems, OLAP, query and reporting tools, data mining, search engines, and web browsers) that can make implementation more difficult.

Both ERP and CRM can be useful to disseminate knowledge but require some of the other core capabilities (i.e., data warehouse) to be in place. ERP is a process and application that enables organizations to better manage resources and allow the integration of daily operational functions (e.g., enabling human resources, payroll, and financial systems to relate to one another and share data). CRM also is a process and application by which a company maximizes customer information in an effort to increase loyalty and retain a customer's business over its lifetime.

HOW WOULD A UTILITY BENEFIT FROM KNOWLEDGE MANAGEMENT?

Many organizations have moved toward a Knowledge Management environment that addresses the need to improve the usefulness of information. For utilities, the incentives to move to this approach include lost knowledge from downsizing and restructuring; privatization of public and municipal utilities; increased employee-customer contact at all levels; the need to improve efficiencies; emerging contaminants and pathogens; and increased regulation. In addition, the competition for customer relationships, globalization, and the need for increased workflow efficiency also act as stimuli for a move toward a Knowledge Management environment (APQC, 1996, APQC 2000). After researching numerous organizations, APQC (2000) identified six major strategies for Knowledge Management:

- *Knowledge Management as a business strategy* – useful for companies to compete and grow
- *Transfer of knowledge and best practices* – useful for organizations that want to improve operations, products, or services
- *Customer-focused knowledge* – focuses on knowing the customer's needs and preferences, while improving communication and service

- *Personal responsibility for knowledge* – used by companies that realize their employees are their most valuable asset
- *Intellectual asset management* – emphasizes intellectual assets including patents, technologies, and operational and management practices in order to gain a competitive edge
- *Innovation and knowledge creation* – emphasizes basic applied research and development

Each case study in the literature consisted of varying levels of information, including the challenge the organization faced, the old approach to the problem, and what solution they implemented. In addition, case studies documented the functionality of the system, the costs and benefits of the approach, and whether the implementation was successful. Table 3.1 lists utilities that used Knowledge Management tools successfully. Appendix A shows the detailed results from this review.

Table 3.1

Review of Case Studies Found in the Literature

Utility	Knowledge Management Tool
Columbus Water Works	Data Warehousing
Tennessee Valley Authority	Data Warehousing
U.S. Geological Survey	Data Warehousing
Essex and Suffolk Water	Records & Document Management and Imaging
Yorkshire Water	Records & Document Management and Imaging
South East Water Limited	Records & Document Management and Imaging
Ontario Hydro	Records & Document Management and Imaging
Omaha Public Power District	Records & Document Management and Imaging
Baltimore Gas and Electric	Records & Document Management and Imaging
Integral Energy	Records & Document Management and Imaging
Public Service Electric and Gas	Records & Document Management and Imaging
OTAY Water District	Records & Document Management and Imaging
Furnas Electric	Records & Document Management and Imaging
Gulf Power Company	Records & Document Management and Imaging
Sydney Water	Workflow/Remote Access
General Electric	Workflow/Remote Access
Wessex Water Services Ltd.	Customer Service/Billing
Three Valleys Water	Customer Service/Billing
Michigan Consolidated Gas Company	Customer Service/Billing
Georgia Power Company	Other (GIS)
GPU Energy	Other (GIS)
Niagara Mohawk Power Corporation	Other (GIS)
Northeastern States Power Company	Other (GIS)
Pennsylvania Power and Light Company	Other (GIS)
Pinellas County Utilities	Other (ORACLE and GIS)

In general, the literature case studies demonstrated that utilities are working toward increased efficiency and cost reduction while trying to increase customer support. This has been achieved primarily through the use of data warehouse, record/document management and imaging, and GIS. While portals, OLAP, and data mining are very important Knowledge Management tools, the use of these more traditional Knowledge Management tools was not observed in these case studies.

WHAT ARE THE LIMITATIONS OF KNOWLEDGE MANAGEMENT?

Initiation of any new technology is a serious undertaking that requires scheduling and prioritization of requirements. Some of the aspects to consider prior to implementation of a KMS include:

- It cannot replace the development of good fundamental standards, documentation, and organization.
- Utility size needs to be considered when implementing a KMS as it may not be beneficial to smaller utilities.
- The actual implementation of a KMS is initially costly. It can be costly to implement comprehensive information systems and sometimes difficult to quantify returns on investment for these systems. Implementation of systems is often far in excess of the costs of system maintenance because of the requirements to convert legacy systems to new technologies. However, some returns on investment, like the ability to produce compliance reports to facilitate interactions with regulators, may occur very quickly. Improvements in efficiencies and reduction in costs will take longer but can be easily predicted and demonstrated. Some of these improvements will include staff reductions and dramatic improvements in efficiencies. Areas likely to be affected first through increases in efficiency include maintenance, operation, and customer support.
- A utility should clearly understand its requirements and the available options for addressing those requirements. A comprehensive requirements analysis is necessary to define the information requirements, mode of interaction, and obstacles to both data access and transfer of pertinent data throughout utilities.

- Knowledge Management implementation depends on a change in culture and use of data and information within an organization. The installation of these tools in the absence of a change in the conduct of business will not have a significant effect, and in some cases may result in poorer performance. It also is essential to designate roles within the organization and focus on organizational structure including identifying specific roles essential to facilitate the generation, sharing, and management of knowledge. Roles should include people who create knowledge, who act as the Knowledge Management facilitators, and those who actually apply the knowledge (APQC 2000). The creators of knowledge include external sources, publishers, analysts, and network members. For example, typical data warehouse analyst roles could include a project manager, a warehouse architect, a warehouse administrator, and a data administrator. Roles within the KM infrastructure include publishing coordinators, knowledge organizers and architects, Information Technology, and perhaps help desk attendants. The final roles of those who apply knowledge include teams who are led by full-time knowledge managers. Roles for managing information within a utility based on our case studies are discussed in detail in the Conclusions section. In addition, each limitation to Knowledge Management is discussed in detail in the Conclusions and Recommendations section.

CASE STUDIES

A significant amount of water quality data are generated by drinking water utilities. It is clear from the case study meetings that the utilities share many common perspectives. They perceive information as an important resource that is necessary for a wide range of purposes. Water quality information is a leading requirement for managing operations, maintenance, customer relations, and compliance reporting. Water quality data generally includes all those components that relate to the sources, operations, and maintenance of the system. Specific examples of water quality-related data includes:

- Monitoring results
- Inline data (SCADA)

- Operational water quality evaluations (usually done by operators)
- Flow rates
- Billing information
- Maintenance information
- Mailing lists
- Customer complaints
- Engineering designs
- "As built" engineering drawings
- System infrastructure
- Permits
- Planning documents

All of these types of information are considered either direct water quality information or are useful, if not critical, for consideration of water quality issues. Some of these sources are surprising, e.g., the kind of water quality data that is obtained from mailing lists. The type of enterprise located at a particular address is often critical information in backflow prevention or in diagnosing water problems. Similarly, billing systems are often the best source of data related to water consumption.

All utilities interviewed agreed that water quality data are emerging as important resources as regulations become more complex and economies are explored within each utility to increase quality with limited resources. Information continues to gain in value as it is used to both fine tune processes within the utility and to meet expanding requirements for complex data and information aggregates. Due to this continued increase in importance and the realization that proper management and analysis of data can result in significant reduction in costs and increases in water quality, utilities are beginning to implement advanced IT to support data access and use.

There is a common set of general information requirements that are consistent among all the utilities. Basic information processing, access and analysis needs include:

- Ability to gather and track samples accurately
- Capturing measurements and observations made in the field and the distribution system
- Processing samples and results through utility and contract laboratories
- Performing quality assurance and data documentation
- Accessing, processing, and summarizing continuous data gathered by SCADA systems
- Tracking observations and notes made by operators
- Managing both scheduled and emergency maintenance
- Managing documents
- Updating and tracking changes to engineering documents
- Mobilizing and delivering documentation related to maintenance and operations
- Developing and maintaining maps of infrastructure, sampling and water quality
- Tracking complaints and issues
- Communicating activities, like scheduled maintenance, resolution of issues, and results of sampling, between departments
- Accessing these disparate information types via online information systems
- Developing a directory of available information and utility information experts
- Maintaining a record of employee certifications and licenses
- Reporting of data for both internal and compliance requirements

Regardless of size and sophistication, all of the utilities studied require these common capabilities and would like to integrate as much of this information as possible. However, none of the utilities studied had succeeded in integrating these resources into a single system, or even into a loosely knit system that could be accessed easily from a single information system. Consequently, there is redundancy in the information maintained in the systems. Frequently, the needed data are available online, but without extensive training, the information cannot be accessed easily or used outside of the primary purpose. The failure to share information results in missed opportunities, redundant data collection, and tension between offices within utilities. The biggest challenges seem to be the integration of pertinent information into an information repository that is readily available and navigable to the majority of the employees within a utility.

The ability of the utility to meet these needs differs dramatically and appears to on each utility's workplace culture. Specific cultural areas that appear to have a major effect on the success of information processing include:

- Utility management
- Communications within the utility
- Interrelationships between departments
- Outside information requirements such as city IT regulations that must be followed by the utility
- Resources committed to IT
- Formality used in developing information systems and infrastructure available to support and access the information system

The utilities studied also exhibited a wide range of strategies for implementing information systems. Most exhibited a combination of centralized and distributed information systems. The centralized systems are in many forms, but are dominated either by legacy systems or new, complex systems implemented by a central IT department either within the utility or within a city. Many of these systems have been implemented to meet general needs and they often do not meet the specific needs of the drinking water utility. Many of the central IT groups are caught in the awkward position of being required to support systems chosen by committees and constrained by city or department standards versus meeting the needs of their utility clients. While top-down requirements may be effective for integrating data across many departments, it often ends up frustrating the users whose needs may not be met effectively. This situation results in the users being alienated from the IT department that is supposed to support them.

Often the best and most expedient solution is for the technical person at the bench level to create a simple information system that meets specific needs without considering the central standards. When this occurs, immediate needs are met but the resulting information is not easily located or accessed by other users. The data are effectively used within a single functional group (department, division, or group) but are not generally available to other groups except through special arrangements. Eventually, other people and departments that need these data find them and then make the data available, either by direct access over a local area network, or by making copies of the original file. This sometimes results in multiple copies that may or may not be updated when changes are made. In addition, secondary users will not find pertinent information resources unless they know whom to ask and are persistent in their pursuit.

The project team found that the focus group meetings were often forums for people who had been working together for many years in the same organization to find out about the availability of data. This observation led to a consensus that an online directory of information resources and utility experts able to help others access and understand data would be helpful.

The extent to which the utilities are adopting IT depends on the size of the utility and the volume of water produced for consumers. All the utilities studied are in the process of evaluating IT upgrades. The areas of interest are consistent and include:

- SCADA upgrades
- Laboratory Information Management Systems
- Maintenance Management Systems
- Geographic Information Systems (GIS)
- Customer relations and tracking systems
- An Enterprise-wide Information System to tie these other systems together

The project team also identified consistent needs for regulatory tracking and records management systems. Most of the functional requirements for these capabilities already are being met through purchased systems, individually generated spreadsheets of databases, or paper systems. In some cases, the functional requirements are not being met, which results in significant costs and potential liabilities to the utilities. For example, the consistent problem with maintaining "as-built" drawings is being handled with varying degrees of success, but has been identified by all utilities as a serious challenge, if not an outright deficiency. When "as-builts" are not maintained, crews often are dispatched without knowledge of the exact location or infrastructure needing repair.

In some utilities, proprietary systems also pose a problem. Many of the SCADA systems are proprietary and difficult to integrate into an enterprise-wide system. These problems can be resolved using export functions from existing systems, but are better solved through the use of SCADA systems that communicate easily with standard database products using ODBC and other standard technologies.

In medium-sized utilities, many sophisticated information systems have either been purchased or developed in-house. These systems sometimes do not meet their original objectives. Most of these appeared to have failed because of one of four major issues:

- Lack of a robust requirements analysis
- Attempt to solve all problems with a single system
- Lack of maintenance of the system or information in the system
- Oversell on the part of the vendor

Any system implemented by a utility should have a focused requirements analysis performed. All stakeholders should carefully document and review the requirements. They should then use the requirements document as a tool to communicate the objectives of the system to both the utility and possible vendors and consultants. They also should use the requirements to guide the implementation and long-term maintenance of the system

Utilities should not try to solve all problems with a single system. Often when a utility is considering a major expenditure for software, it attempts to meet too many needs with a single purchase. This often results in the system not meeting its intended uses or solving the extended issues. Once the system is installed, resources may not have been allocated to maintain both the system and the data in it. This results in a failure although the purchased solution may have all the required functionality.

It also is important to research the vendors. Vendors often will sell whatever the utility says it needs. Without a careful assessment of the needs and an evaluation of the software's ability to meet these specific needs, expectations may be too high. Once the sale is complete, the utility may not be in a realistic position to litigate or to insist on holding the vendor to its original commitments. Forcing a vendor to address requirements in writing can alleviate this problem.

PILOT

The Knowledge Management pilot attempts to integrate water quality to demonstrate the capability of utilizing knowledge from a variety of sources and information types as part of a decision-support system. The project team reviewed applications of vendor components and in-house products for the development of the pilot. We were interested both in developing tools to provide some functionality toward the project goal, and delivering information aggregates to a subset of customers of the Tucson Water Department, the institution hosting the developed system. Such tools enable information sharing both within a utility and with outside sources. The availability of the data to outside users must include the development of a proper security system. Some of the functions included in the pilot follow:

- Enable the user to summarize and aggregate information, and create simple tables and graphics
- Develop a predefined set of information for the screen comprised of daily reports, live graphics, and alerts
- Enable the user to perform both programmed and ad hoc searches for information
- Enable users to search for relevant documents from full text indexing of the entire shared file system

Since this was a pilot, the project team provided a restricted set of capabilities demonstrating the functionality of Knowledge Management. To expedite the production of the pilot system, we based the core infrastructure on Open Source (source code freely available) technology, except for the database systems, which use Oracle for the warehouse. The key technology is the application server, ZOPE, and a powerful scripting language, Python. ZOPE is an object-oriented system that can be used to rapidly design web-based applications, especially when joining to database systems. Using Open Source systems allowed for cost savings, as well as the ability to utilize third-party extensions. These extensions give improved graphing capability and document access for indexing, as well as the ability to communicate with e-mail systems. In a real-life implementation, use of off-the-shelf products may be required, so a complete integration procedure will be necessary. This may require significant resources, but will result in a complete and easy-to-use system. This integration may require the service of a consultant as well. It is imperative that an organization recognize that this route can be cost-prohibitive for smaller utilities.

Conceptually, the system is comprised of multiple clients communicating with a server over the utility intranet. The server provides a number of services by way of a web server. The services (as depicted in Figure 3.5) include:

- Personalization: User's settings, preferences, screen layout, etc. are stored in an object database and used to personalize output every time a request for other services is made.
- Document Repository: Documents are delivered from a shared file server in response to document search requests.
- GIS Layers: Requested layers of spatial data such as water quality zones, well pumpage and sampling points are delivered on demand.
- Data Services: Requests for data from a variety of sources such as water quality database, SCADA (Supervisory Control and Data Acquisition) systems, customer service database and well database are handled on demand.

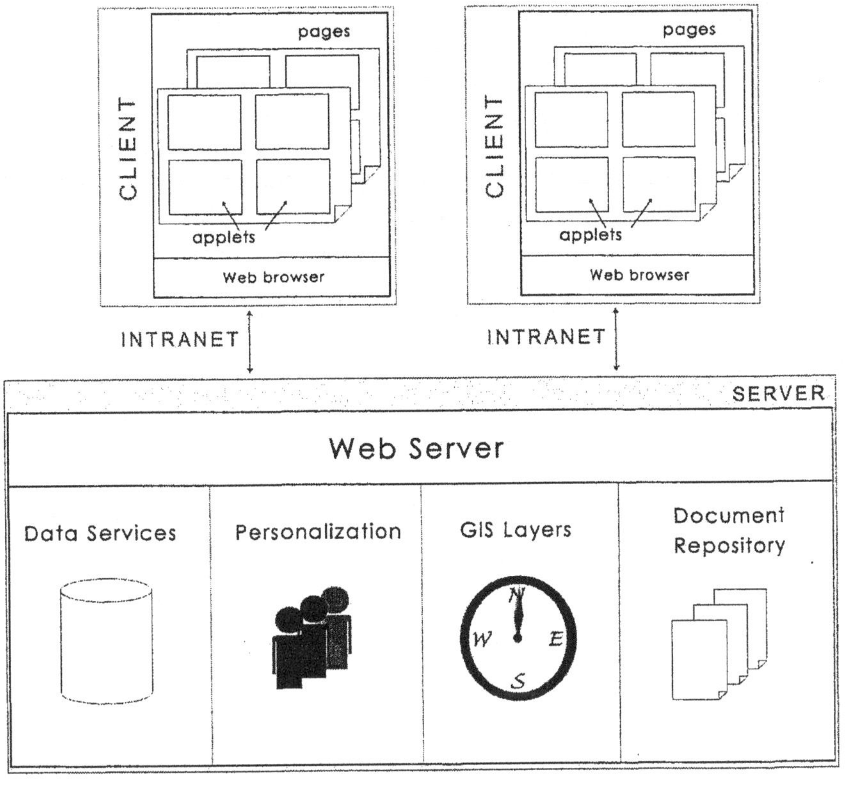

Figure 3.5 Conceptual design of the system

The clients are simply web browsers on each user's desktop. When logged onto the system, a user's screen may have a number of "pages" each containing "applets". Pages are containers for applets and are meant to organize applets together. Each applet is a small application that conveys information and/or facilitates access to information such as a chart, map or document search applet. Detailed description of page/applet paradigm is given in upcoming sections.

The pilot is set up so users can customize what they see and how they relate it to the information available in all underlying databases. The user logs on to a portal that helps him or her create a window to the available information (Figure 3.6). The user can accomplish this in two different ways. First, a set of standard interfaces that display a wealth of information in preprogrammed formats is supplied. The user can choose and reset these during any session. Second, the user can choose from a list of optional menu items (applets). Each item represents a small window of information that can be placed on the screen. Once the user likes the setup and content on the screen, he or she can easily call the configuration in a custom user profile or set it as the default presentation. Independent of what options are chosen, the user can add new applets to the screen, or modify existing applets, at any time.

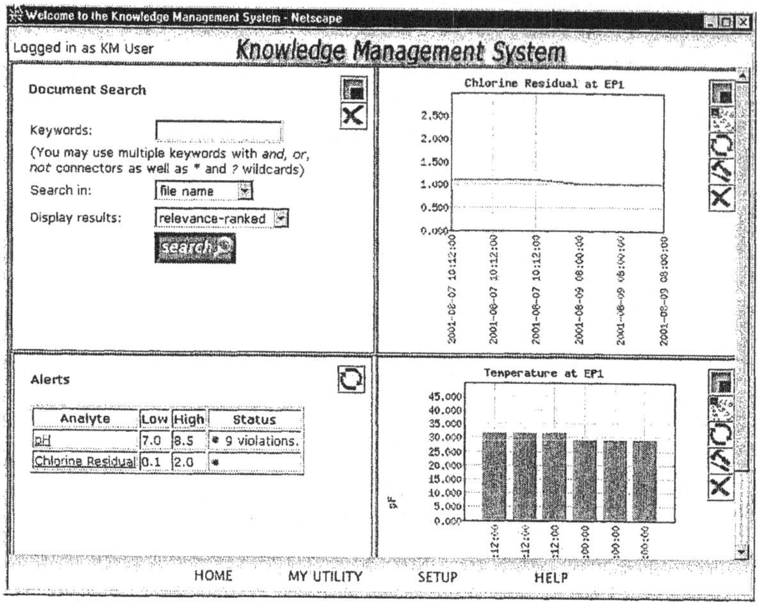

Figure 3.6 “My Utility” Screen

When designing a custom profile, the user may add applets through the data wizards, with the exception of the document search tool. The user can pick a profile or rearrange a screen based on which applet is displayed at what location. After the user chooses a profile, the “My Utility” screen (Figure 3.6) is generated by that profile. This screen is fully configurable by the user through the selection of applets. Applets have been designed in several forms, including:

- **Chart.** This displays the data in a chart format, such as a line chart, bar graph, area plot, or pie chart. The user has the option to create this chart in 2D or 3D, along with many other options. For example, the user can choose titles and labels, axis scales, time scales, and legend. The user also can define the grid type, and can display or hide the axis labels, depending on the type of chart. For pie charts and area charts, the user can select the colors desired for the presentation. Currently, there are default scales for the user, but this may be modified in the future to allow manual selection of the scale.

On the "My Utility" screen, there are a number of buttons for each chart. A "tools" button opens the Chart Options wizard, which walks the user through configuration of the applet. Figure 3.7 shows an example of a screen within the Chart Options wizard. As the user selects from the available options, a preview of the resulting chart is dynamically generated. The user may continue through multiple screens, or may return to the "My Utility" screen at any point.

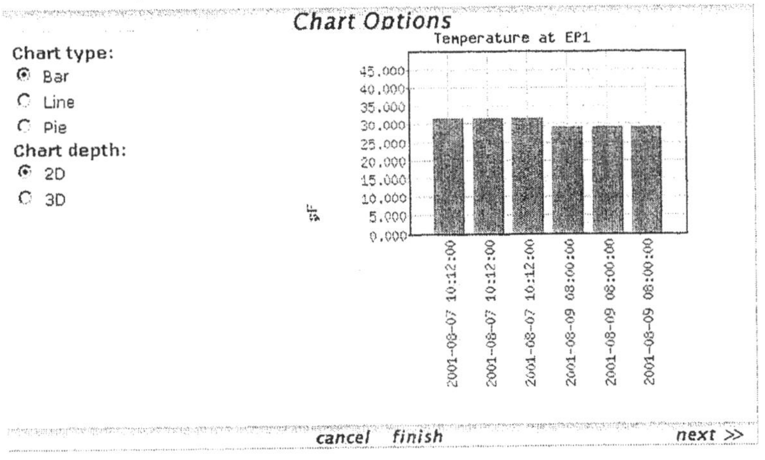

Figure 3.7 Chart Options Wizard

The user also may use the "drill down" button to display the data behind a chart in tabular form. In Figure 3.8, the user clicks on this button to open a new window with more detailed information.

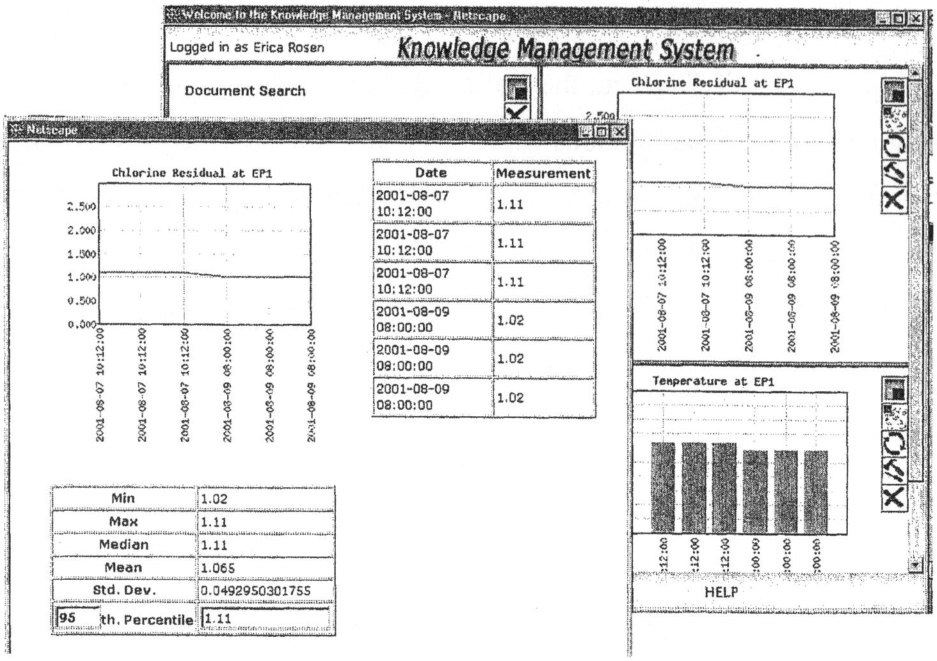

Figure 3.8 Displaying Detailed Information with a Chart's Source Data

- **Map.** Map applet serves two purposes. Like any map, it provides a visualization of spatial data. In the current implementation of the pilot system, for example, the boundaries of water quality zones and the location of sampling points and wells are plotted on a map. In Figure 3.9, blue dots represent the sampling points. Red dots represent wells that are currently pumping while the size of the dot conveys how much water a well pumps in relation to others, i.e. the larger the dot, the larger its pumpage. Wells that are not pumping are marked with green dots, which are not very noticeable at current zoom level. The map can be zoomed in and out or panned around. As Tucson already has in place a full-featured web-based GIS interface to water quality data, the pilot focuses on giving the manager a quick overview of a selected data attribute averaged in the pressure zones of the city as well as the additional information on pumpage activities.

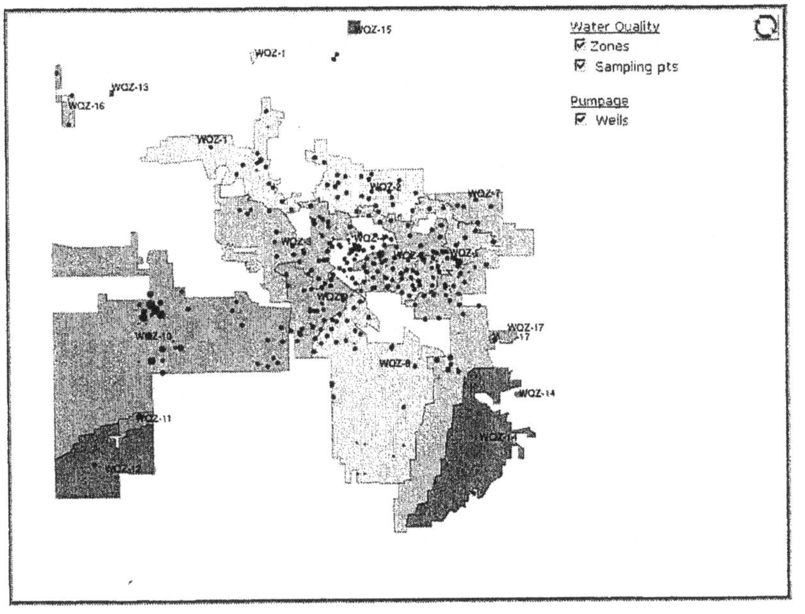

Figure 3.9 Map Applet

- **Document Search.** The Document Search tool is able to catalog files (i.e., .doc, .wpd, .xls, .csv, and ppt, .pdf, and groupwise e-mail) and search by file name or content to obtain relevant documents. Searching mechanisms support options such as AND, OR, NOT, and other operands and/or multiple keywords.
- **Alerts.** A mechanism will be available for the user to set up alerts. For example, the user may set alerts based on data thresholds (i.e., "Once chlorine residual gets to X level, send alert"). This alert will be textural and instantaneous but also will be stored in a file for alert notifications.

The pilot system was made available to staff at Tucson Water and we had 13 staff members set up account in the system for review and from those 7 questionnaires were returned with feedback on the system. All responses were very positive and the primary weaknesses all dealt with the desire to add additional data sources. The primary next step identified is for cost estimates for a production system. The development costs depend upon:

- What software platforms are required for implementation? The use of Open Source technology greatly reduced cost, but is often not accepted by organizational standards. The use of Commercial Off The Shelf (COTS) software brings the additional cost in that it is a collection of tools probably not covered by any one or two COTS products.

- What information infrastructure is in place? The selection of Tucson as a trial site was chosen because they have already integrated many data sources into a centralized warehouse.
- How many different data sources are desired for integration? This is potentially an open-ended requirement. As indicated in our questionnaires received, practically every individual was envisioning the addition of yet another data source.

Excluding software costs, it is expected that the entry point into development of a baseline system would start around $75,000 to $100,000 for development costs.

CHAPTER 4
SUMMARY AND CONCLUSIONS

RELEVANCE TO THE DRINKING WATER COMMUNITY

Information is a critical resource that is currently handled most effectively by only a few drinking water utilities. While all daily operations may be run efficiently within each utility division, the result is an enormous amount of data and information used mainly for primary purposes. Typically, utilities are inundated with samples and reporting requirements that require effective management and data analysis. Individual utilities have addressed this need for improved information management and use in a fragmented manner. While many utilities have purchased and installed LIMS, GIS, SCADA, and hydrological models, most have not considered the overall workflow efficiency among all of the systems. Consequently, data transfers may be rendered incompatible or too complex, resulting in limited use of systems among divisions. The organization, maintenance, aggregation, dissemination, and application of the integrated information from these systems needs to be utilized in an effective manner across the entire enterprise to support the operation and maintenance of the utility. This project documented the current use of information in a utility, as well as how it should be used. It also identified the processes required to make information useful to a wide audience, specifically the senior management within a utility. Ultimately, this project established the need for the development of an Information Bureau within utilities to facilitate this entire process.

While the concept of an Information Bureau is useful for utilities, this project also documented the information needs and requirements common to different utilities, as well as some that are unique. The process used for this, based on standard requirements analyses, should allow the drinking water community to have the opportunity to evaluate the effectiveness of new approaches without first implementing technologies that can be quite costly. Many problems drinking water utilities have encountered when implementing such also may be avoided through a good comprehensive requirements analysis. The outputs for the requirements analysis also can include detailed requirements for requests for proposals. This will ensure that the requirements are clearly articulated and that bidders are responding to the real needs.

In addition, the development of a knowledge system pilot may act as guide for other utilities to develop an enterprise-wide information system. Both Tucson Water and Des Moines Water Works (DMWW), under EPA's Environmental Monitoring for Public Access and Community Tracking (EMPACT) program, are examples of utilities making water quality data available via web-enabled water quality information systems. Both projects include interactive realtime information, integrating daily collection, analysis, posting, and updating of drinking water quality data. These examples demonstrate some of the capabilities of the pilot but are oriented more to the public. The audience for the pilot is the utility staff, which is likely to be more sophisticated in their understanding of data and have needs for both aggregated and raw data .

This project has demonstrated how a requirements analysis can be done at a utility with limited time and resources. Appendix B presents a model for the information covered in a requirements analysis. The appendix also presents detailed summaries of the requirements, barriers, and recommended solutions. These can be used effectively by other utilities to anticipate requirements, barriers, and potential solutions. This project has identified and categorized the required building blocks that need to be considered by drinking water utilities if they are to improve their use of information to upgrade operations, compliance reporting, and decision support. If the functionality of the pilot is appropriate and meets the needs of utilities, this project has developed a model that can be applied by other utilities.

WHAT IS KNOWLEDGE MANAGEMENT?

There is a broad range opinions regarding exactly what Knowledge Management is. American Productivity and Quality Center (1996) and Firestone (1998b) view Knowledge Management as strategies and processes that are capable of identifying, capturing and maintaining an organization's knowledge. SAS Institute defines Knowledge Management as a system-based approach that is capable of capturing and leveraging institutional memory. Butler (1999) defines Knowledge Management as an integrated approach comprised of processes, technologies, attitudes, and systems that promotes and provides for adaptation and survival of the organization in a rapidly changing environmental workplace. Although there are discrepancies in the definitions, the main goal of Knowledge Management is to identify, capture, manage, analyze, and extract knowledge from various sources within an organization and place it into a structured, organized format (that may need to be refined over time) that can be shared. Knowledge Management needs underlying data, standards and organization.

WHAT IS A KNOWLEDGE MANAGEMENT SYSTEM?

A "Knowledge Management System" (KMS) is not a package that can be purchased like a statistical or accounting package. A KMS, which will provide organized access to knowledge, must be developed and applied as a solution for a particular organization. It consists of many components selected for the particular capabilities that can be integrated to meet specific requirements. It is a solution based on many tools and is not an off-the-shelf package. While KM does not need to be computerized, many formal KMSs are. KMSs often interact with databases, e-mail, document management systems, and operating systems directories. The support technologies that assist with Knowledge Management include data warehousing, records and document management, data mining, OLAP, portals, enterprise resource planning, and customer relationship management.

CULTURAL CHANGE

Knowledge Management implementations are not only contingent on the installation of new hardware and software, but also depend on a change in culture and use of data and information. The installation of these tools in the absence of a change in the conduct of business will not have a significant effect, and in some cases may result in poorer performance. Knowledge Management requires a significant change in a utility's organizational culture to be successful. This cultural change is often more difficult than the investment and challenge of gathering and organizing all of the required information. A utility needs to commit to become an information-based organization. System implementation needs to include extensive training in the specific application of the system especially for middle and upper managers.

COMMON UTILITY NEEDS AND BARRIERS

A major conclusion of this project is that both content and function information requirements are similar to a degree for all utilities studied. Each utility has unique requirements but a core set seems to be consistent. Basicneeds for water quality information processing, access and analysis include:

- Ability to gather and track samples accurately
- Capturing measurements and observations made in the watershed, source, treatment facility and in the distribution system
- Processing samples and results through utility and contract laboratories
- Performing quality assurance and data documentation
- Accessing, processing, and summarizing continuous data gathered by on-line and independent field instruments
- Tracking observations and notes made by operators
- Managing both scheduled and emergency maintenance
- Managing documents
- Updating and tracking changes to engineering documents
- Mobilizing and delivering documentation related to maintenance and operations including operating manuals and standard operating procedures
- Developing and maintaining maps of infrastructure, sampling, and water quality
- Tracking complaints and issues
- Communicating activities among departments (like scheduled maintenance, resolution of issues, and results of sampling)
- Accessing disparate information types via online information systems
- Providing a directory of available information and utility information experts
- Maintaining employee certifications and licenses
- Reporting of data for both internal and compliance requirements

- Development of trending analysis of data
- Integration with hydraulic models

Most utilities exhibited a range of strategies for implementing information systems. Solutions covered a wide range of approaches including both centralized and distributed information systems. Most of the information gathering and management appeared to be compartmentalized within a single department. Most of the initial data gathering occurred in individual databases generated by a particular group or department. In the cases where these data were then centralized, they usually were transferred and loaded by either the group that originally collected the data or by the IT group. While there were a number of instances of integrating the data into a centralized system automatically, this was the exception rather than the rule. Many of the systems that were observed during high-level requirements analyses met broad needs defined by vendors but did not meet the specific needs of the utility. The project team noted examples of these shortcomings for LIMS, maintenance management, and many of the systems imposed on the utilities that were municipal departments.

The project team identified a number of consistent barriers to the management, dissemination and use of information resources within utilities during the case studies. The observations and barriers identified at the utilities fell into eight overall categories including 1) Directories, 2) Data Availability, 3) Data Quality, 4) Communications, 5) Resources, 6) Data Processing Tools, 7) Standards, and 8) Training. A brief description of each is provided below.

1. **Directories.** This was a consistent issue observed at all utilities. Data are stove-piped within functional organizations, and data stores tend to be recreated by multiple individuals without the knowledge that someone has already developed it. It was clear that there is a lack of understanding of what information is available, how to locate and access it, and what experts were available to give the data context. Many people who worked together in the same organization for several years did not know the appropriate data contacts, how to access data across departments easily, and would often have to meet to discuss their desires for data accessibility.

2. **Data Availability**. There are a broad range of data types and sources within a utility. A lot of the data availability issues arise from the tendencies for utilities to use "people centric" and "paper centric" systems. These systems often have data that are valuable utility wide stored on individual and/or departmental servers (stove piped databases – only available within a department or group). Historical data may be available in paper format only or stored in files with obsolete formats, that are not easily accessible to aggregation tools.
3. **Data Quality.** While utilities are very good at capturing raw data they are not consistent in developing documentation regarding data quality and processing so it is often difficult to integrate data collected for other purposes. Typically, problems exist with data aggregation as a result of the lack of standards both within the utility and within the drinking water industry. Lack of standards often results in inappropriate data being aggregated. Lack of documentation often results in data not being used or not being used properly.
4. **Communications.** A major issue identified was that many potential data users within a utility are not aware of a significant amount of the available data. In general, naming conventions are inconsistent between groups within the utilities. In addition, sharing data between departments is not an easy task and there is limited communications within the utilities between divisions. It also appears there is reluctance to communicate electronically in some groups within the utilities.
5. **Resources.** Most utilities' IT departments are understaffed for the goals and requirements including development, maintenance, help desk, data access, analysis and reporting. In addition, data experts within divisions are not recognized and responsibilities are not formalized.
6. **Data Processing Tools.** Most utilities have inadequate data query and subsetting tools. There is apparent underuse of standard analytical tools for data subsetting analysis and visualization. Often GIS, graphics, report generation, databases and statistics are perceived as difficult to use.
7. **Standards.** Often the quickest solution for most departments within utilities is to create a simple information system that meets their specific needs without considering any central standards. While this may serve the immediate purpose, it causes

more problems as other departments eventually need to locate and use these data. The lack of standards, or inconsistent ones, was consistently noted at utilities for many critical elements including naming conventions for locations and facilities, data tables and fields (entities and attributes), data directories, and software.

8. **Training.** Many utilities are doing very well with training and support on the standard desktop suite. However, training and follow through is not always adequate for advanced systems and software (e.g., GIS, SCADA, Database, Maintenance management, Graphics, Statistics, Modeling).
9. **Cultural Change.** As previously mentioned, organizations have an inherent inertia when it comes to change. The resistance to change by employees must be recognized and accommodated.

Knowledge Management can be a major tool to help resolve some of these issues. The resulting improvements in information access and use also must result in improvements in operating efficiencies and water quality. The resulting system needs to allow a wide array of users within a utility access to data, information, and knowledge that is maintained in a variety of forms and systems.

RECOMMENDATIONS

It is critical to understand that the transition to a knowledge-based strategy is an expensive undertaking that must include proper hardware, software, personnel, policies, and corporate commitment. Instituting a KMS needs to begin with a clear understanding of requirements and priorities.

Reluctance to change the way that data and information have been handled historically is one of the biggest challenges for utilities when implementing a new system. While many employees are willing to move into an electronic mode for standard business operations, others are reluctant. This can be resolved through training, edict (to a limited extent) and attrition.

1. **Information Bureau.** A drinking water utility is a complex organization requiring, at a minimum, some infrastructure to handle both computers and communications. This was clear even for the smallest utility. Drinking water utilities have developed Information Technology groups or expertise. In the case of small utilities, this function may be served by a contractor. In small municipalities, there may be a small Information Technology staff that serves a number of departments. As the size of the utility increases and the complexity of treatment, sources, and distribution increase, the complexity of the infrastructure increases at least proportionally. Therefore, each of these utilities requires a staff capable of maintaining, operating, and continuously improving the computer and communications infrastructure. In addition, given the importance of computers in managing data and communications, there is an ongoing need to maintain computers, support and upgrade computers, and support both software and users. This should be the domain of the Information Services group (also called the MIS or the IT group). This group also should be responsible for the development, implementation, and maintenance of large utility-wide information systems. The group members should possess mostly Information Technology skills and education.

   These professionals often are called upon to support data processing, analysis, and interpretation, but it is not their primary responsibility. These IT professionals frequently are overwhelmed with the maintenance and operations of the information systems, leaving them little opportunity for knowing the information content of the system or the correct use of the data resulting in little support for locating and using the available information. This is paradoxical since the whole purpose of the information system is to expedite the location, access, and analysis of the data. This failure is not due to any shortcomings on the part of the IT staff, but rather in a lack of clarity of their responsibilities and priorities. The IT departments are regularly understaffed and cannot meet their primary responsibilities, which are usually to maintain the integrity and functionality of the hardware, software, and network.

This problem can be solved by the development of a team within each utility that focuses on the data. Their mission needs to be an understanding of the available data and a passion for helping users access and analyze these data. This group will be referred to hereafter as the Information Bureau.

The Information Bureau concept is not meant to be an extension of the Information Services group at a utility. These are two completely different roles that require an entirely different set of skills. These requirements revolve around locating appropriate information, accessing, using it correctly, finding expertise, understanding how the information was collected, and understanding its appropriate use. These requirements are best dispatched by subject matter experts, not by Information Technology personnel. At each utility studied there are a few technical people who know the available data better than most. Many people within the utility know to whom to go for help in finding, accessing, and using data. Their position descriptions never include that they support the access, downloading, and analysis of data but they should be utilized to maximize the applicability of data in supporting good decision processes.

Drinking water utilities should formalize the role of these critical people and clearly identify them to the rest of the staff. These individuals should be properly trained and presented with additional tools to help them access and analyze data. The organization of the Information Bureau within each utility must be worked out to meet the particular organization and management structure of that utility. The following specific recommendation should be applicable at all utilities.

A. The Information Bureau must be led by a senior person able to act as an advocate for the group with management and other senior managers within the organization. The senior person should be chosen based on a zeal for supporting others and for ensuring that decisions are made with the best information available to the decision makers.
B. The Information Bureau should be independent but closely associated with the IT group. These requirements should not compete with the requirements to maintain the systems, support the users, and deal with infrastructure emergencies. These

infrastructure issues regularly supercede locating, distributing, and analyzing data. These issues usually are the primary mission and responsibility of the IT group. Any new resources allocated to the IT groups are likely to be, in part, allocated to support the primary mission and not focused on the intended roles of the Information Bureau.

C. The Information Bureau should have a number of information experts who are dedicated to specific departments but still answer to the Information Bureau Director. In many cases, the ideal people for these positions are already in the departments that they will support. The formalization of the Information Bureau will result in strong connections among the departments and will facilitate the exchange of information because experts in each department will be in communications with experts in all the other departments.

D. The Information Bureau must have a budget for training, software, and improved hardware. These people now do the job using whatever means they have available, which often are insufficient or inefficient.

E. The initiation of the Information Bureau should include an in-depth requirements analysis with participation of both the Information Bureau and the IT department.

2. **Directories: The development of a comprehensive data directory.** Utilities should build and maintain a data and information directory. The establishment of an online directory of information resources, coupled with utility experts, would serve as a great asset to any organization. The Information Bureau should be responsible for maintaining this directory. The data directory should point to the pertinent data available within the utility. It should indicate who maintains the data, where it is located, and where pertinent documentation about the data resides. The data directory also should include a comprehensive data dictionary that indicates the meaning of any data element (also called attributes, variables, or fields) found in the utility database or data warehouse. The data dictionary should document as much information about the field as possible, including the sources of the data, units, methods, etc. that are specific to the data attribute. The data directory also should include tables defining any codes used in the system. In summary, the directory will define where to locate data

and will help the user put the available information into a context that will support proper integration and aggregation of the data.

3. **Quality Assurance**. Quality Assurance (QA) includes thorough evaluations of the data produced or captured. Quality assurance needs to be addressed at every level. QA includes data verification, which is the process that ensures that the numbers and characters were properly input into databases or warehouses. IT needs to be checked each time a new program is developed and each time manual data entry is performed. QA also includes data validation, which is the process of ensuring that the data and information make sense and are in the proper context. This is done by statistical testing and through critical review by subject matter experts. Each time data aggregation is performed, there are risks that the analyses are faulty. Some common mistakes include incorrect units, improper calculations, and incorrect assumptions. A significant risk with computerized systems is that once the data are entered into a database they are incorrectly granted a certain confidence that may be incorrect. Data mining tools are very effective at embedded background quality assurance, including range checking, outlier analysis, and quality control charting. As historical data are migrated and incorporated into enterprise-wide systems, the data must be carefully evaluated from a quality assurance perspective. Detailed documentation regarding the conversion of the data from the original format to the new format must be developed.

   Proper Quality Assurance also includes proper documentation of processes, methods, assumptions, and calculations. These should be cross-referenced in the data directory. Knowledge Management tools like data warehousing can help organize and maintain these documentation files. Utilities should develop and follow quality assurance plans. This may include, but not be limited to, the development of project codes, quality assurance and quality control review, and the standardization of data aggregation. The Information Bureau would be integral at applying tools to perform quality assurance, as well as documenting the resulting data. It would need to work closely with information system programmers and with subject matter experts to ensure that quality control checks were performed and the results documented.

4. **Communications.** Recommendations 1, 2 and 3 above, are the critical first steps in improving communications. The Information Bureau will help communicate what

data are available and will have interdepartmental contacts that can help locate and disseminate desirable data. The development of a data directory will help users know what information is available. It will also help users identify the experts within the utility who can help them apply the information effectively. Formalization of Quality Assurance, as described above, will help users understand the available data and will allow them to evaluate its usefulness for each application. The development of appropriate documentation will ensure that the information is appropriate for its intended uses. Communications can be enhanced through a variety of methods, including developing standards for naming locations, facilities and analytes; linking data dictionaries to directories for data identification; standardizing database entities and attribute names; and developing integrated information systems and records management.

5. **Resources.** It is clear from the requirements analyses that there is an extensive need for additional information processing, analysis, and application. Successful IT departments are now overwhelmed with requests for new systems while the numbers of their technical staff remain constant. The division of Information Technology and Information Bureau into separate functional groups will require some increases in staffing. Many of the Information Bureau personnel will come from existing staff, but the leadership of the Information Bureau may require increases in full-time equivalents.

   It is important for both IT and Information Bureaus to begin documenting the return on investment, including documentation of tangible results, from information systems and from the application of knowledge. Examples include improvement in the productivity of key personnel, in water quality, and in compliance reporting, and reductions in required personnel. Without these tangible improvements, it will be difficult to justify the investment necessary to enhance IT personnel and Information Bureaus, and to implement Knowledge Management.

   Short-term requirements for additional IT personnel could be addressed through contracts, but it is unlikely that the Information Bureau strategy could be implemented through contracts. Contractors could be used to help implement and operate systems. Information Bureau personnel must be familiar with the utility and with personnel in other departments. In summary, possible solutions to the resource problems identified during case studies include increasing full-time equivalents for IT, formalizing responsibilities of experts outside of IT, contracting out development, and some data processing. In addition, five to seven percent of the total staff appears to be an appropriate IT staff size for a drinking water utility.

6. **Use of Data Processing Tools.** The development of standard reports as well as the acquisition of and training in basic software tools would increase knowledge sharing within a utility. Ultimately, embedding these tools in a KMS would be the most efficient method to facilitate communication among systems. At a minimum, this set of tools must include a basic desktop suite of tools and access to the Internet. However, there were some concerns expressed regarding Internet access abuse. While these concerns are understandable, access to the Internet far outweighs the risks. A clear statement of Internet use should be made and then access should be granted. All users should be adequately trained to use these basic tools. Most utilities studied needed improved data management and ad hoc query tools for databases. Some subset of the utility staff should be trained in the design and implementation of databases. This should not be a large percentage of the staff, but key individuals in each department of the utility should be trained in the design, implementation, and maintenance of relational databases. Information Bureau staff would be perfect candidates for this advanced training. These same users also should be trained in the use of data query tools that develop SQL queries. Each utility should have a query tool that can be used to develop and store queries.

   On a different level, utilities should begin to consider formalizing records management. These tools are quite expensive and need to be considered carefully before they are purchased and implemented. Throughout the utility evaluations, it was clear that all utilities, regardless of size, needed records management. Utilities in the United States have become increasingly aware of the critical need for RM systems in the litigation environment. Although it is a sizeable up-front cost, it is inexpensive insurance in the case of litigation.

7. **The Development and Use of Standards.** At each utility studied, there were either no standards or many conflicting standards, which were a major barrier. Standards are necessary for Quality Assurance, data documentation, data processing, and opera-

tions. Unless there are standards, the organization is unable to link data contained in separate information sources pertaining to the same location or analyte. This is an important point that must be addressed if integrated information and knowledge systems are to be implemented. Standards allow data to become information and eventually knowledge. If standards were implemented beyond the organization, information from multiple organizations could more readily provide a review of regions or industry wide.

A list of standards that should be considered includes:

- Naming Conventions: Methods, Analytes, Taxonomy, Sampling Locations
- Database Designs: Tables (Entities), Parameters Measured (Fields, Variables, Columns, Attributes), Relationships, Regulatory Triggers, Goals and Action Levels.
- Content: Formats, Codes
- Documentation

Some of these required standards, including analyte names and documentation, already have been addressed by other industries. Others need to be addressed within a utility or across all utilities. At a minimum, utilities need to standardize naming conventions for locations, attributes, and codes. Failure to address these issues will prevent information integration.

8. **Training.** Most utilities have excellent training in desktop suites, but need in-depth training in other tools. In some cases, communication issues and use of local and wide area networks could be extended, solving a number of problems through consistent training and refresher courses. Some barriers to data access and integration include being intimidated by computer applications or not grasping concepts during the initial training.

   As discussed above, training of a few key personnel in advanced system design, development, and maintenance would be advantageous to the entire utility. Utilities should focus their training needs on requirements identified in a comprehensive requirements analysis.

PILOT

To demonstrate the potential utility of Knowledge Management tools, we developed a pilot software system for deployment in the Tucson Water department. Our case studies have shown that drinking water utilities are still limited in their ability to utilize and share the large amount of data at their disposal. In the case studies, most utilities are good at collecting raw data and storing the information in an electronic format. (D1 level, Figure 3.4). In some cases, the utilities adequately perform Quality Assurance and control, but are not consistent.

The pilot was a portal geared to the senior managers and was an attempt to demonstrate some of the proposed solutions recommended in the case studies. In order to limit the pilot development, we placed the focus on the tools that can utilize the data currently available in Tucson Water. The project team was involved in supporting Tucson Water in developing an integrated water quality information system and a web interface that allows users within and outside the utility to access quality-assured data in near real-time. Tucson Water is currently operating effectively at the D1 (Raw Data) and D2 (Quality Assured Data) levels and has begun developing standardized data at the D3 (Aggregates and Summaries) level that can be generated and viewed in near real time. Some of the D3 data can be generated using both ad hoc and programmed data queries that can then be displayed in tables, graphs and maps. The pilot takes advantage of these existing foundation components and allows users to define the content and format in which they view the underlying data. The community supported by the pilot probably is limited. The readers of this study should envision the application of the approach by any user within a utility, given the customization ability built into the pilot system.

The view of the pilot system is that of a senior manager quickly needing to review the "State of the Utility." For instance, a number of items are available to show the operations and the quality of water being delivered. The pilot draws water quality information from the water quality database, LIMS, SCADA and field data. Future implementations of potential data sources include those databases listed in Table 4.1, which is significant because all components contain important water quality data that must be accessed regularly. It often is difficult for individuals to tie these disparate sources into usable information. The combination of an Information Bureau and a Knowledge Management System can result in a broader understanding of these many sources and how they relate to each other.

Table 4.1

Core Water Quality Information Sources

Water Sources	Backflow Prevention
Field Data	Maintenance Management System Procurements
Operator reports	Billing System
Water Quality Data	Customer Relations
SCADA Data	Budgets
New Developments	Human Resources
Permits	Customers/Mailing lists
Valve Maps	Property records
Capital Improvements	Regulating agency databases

Figure 4.1 represents a typical smaller utility where there is little data shared among systems. The information available to managers largely is contained in a few written reports and supported by verbal communication with supervisors from the various sections of the utility. Figure 4.1 is a depiction of how the utility manager's morning summary is developed.

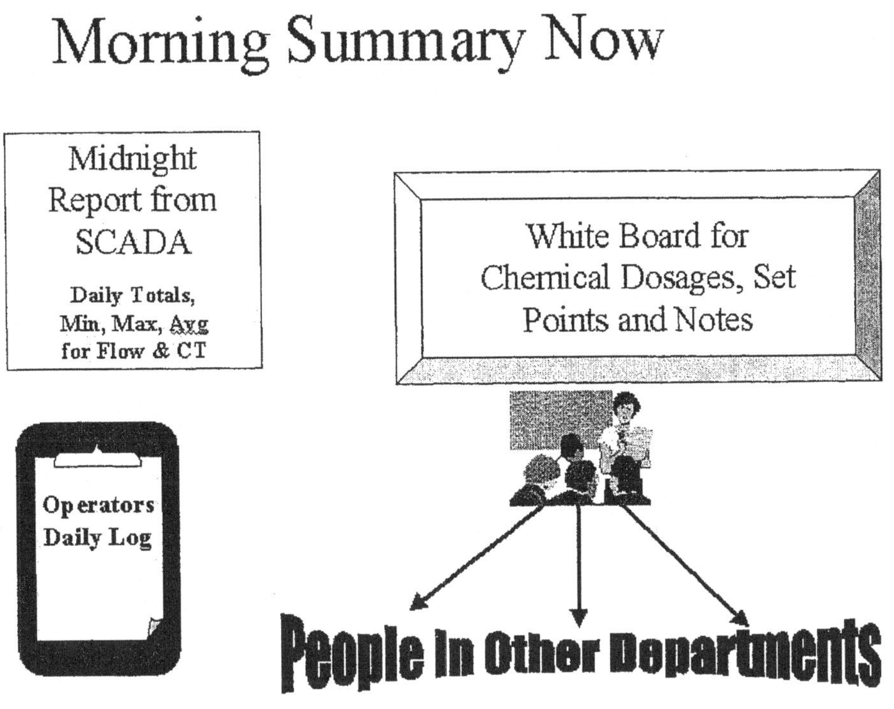

Figure 4.1 Typical Morning Summary in a Smaller Utility

Much of the required information is printed out from individual systems and operator reports are copied or printed and presented to the manager. The manager integrates the information on a white board or other handwritten medium. The manager then calls others in the utility to discuss the current situation and make any required changes and clarifications.

In Figure 4.2, as the utility's information becomes more centralized, more of the information useful to the manager can be automated during its production. At this level, more of the information is available electronically and distributed over the network. There is potential for a common report instead of a stack of seemingly unrelated pieces of paper. However, the staff still needs a great deal of verbal communication with section heads to obtain explanations of any abnormal indicators.

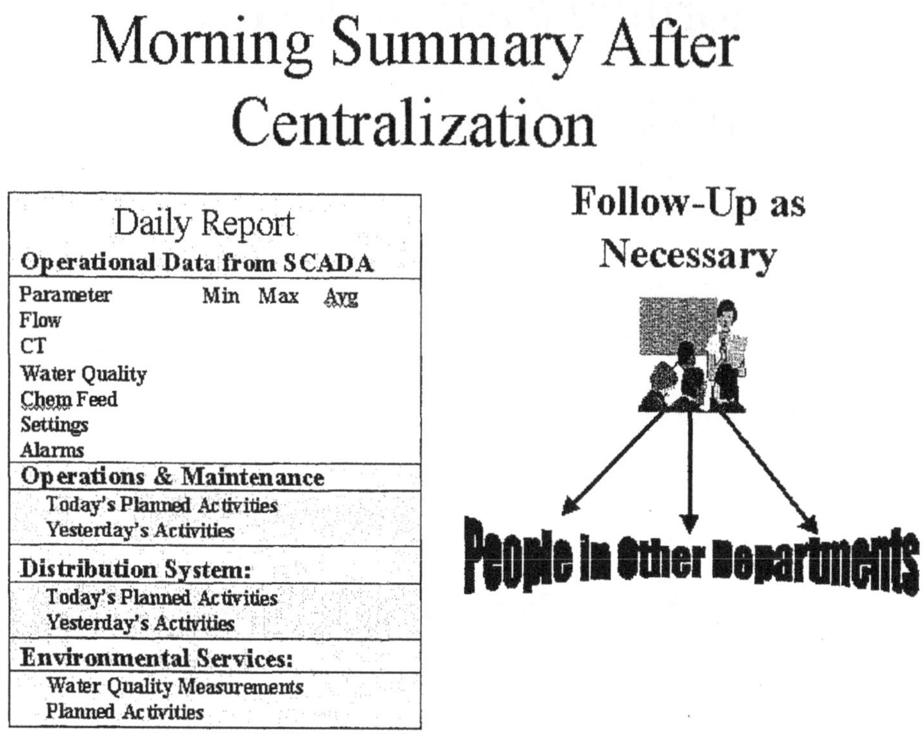

Figure 4.2 Morning Summary after Centralization

The goal with Knowledge Management tools is to build into a set of capabilities based on the centralized data. At the top level, the electronic report to the manager may be very similar to that of the previous figure. The system proposed in Figure 4.3 differs because it is a dynamic system. When the manager sees an abnormal indicator, the first response can now be to "drill down" into the items to view the information in greater depth and detail. This approach allows for the important quick picture of the system as a whole and the ability to view the details of only those aspects important to the current daily activity. This approach does not eliminate the need to hold discussions with supervisors, but these discussions can be more focused on the few details not available or understood by the manager.

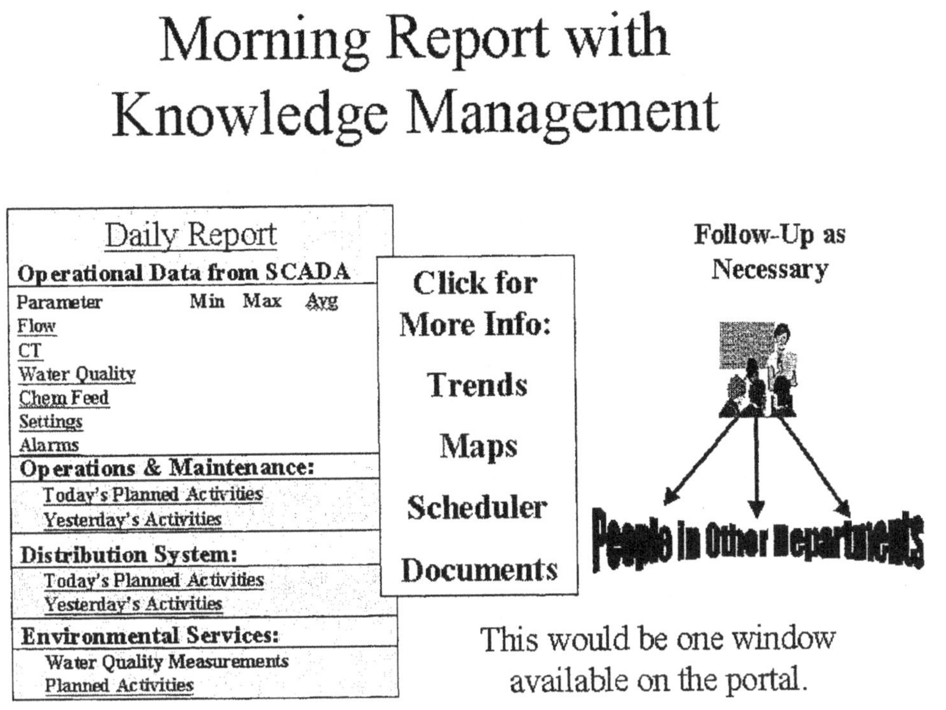

Figure 4.3 The Morning Report with Knowledge Management

Since the state of most utilities was in the area of D1 to D2 (see Figure 3.4), the pilot was focused on this type of data and information. The pilot intended to provide information to a senior management audience who needed information in a quick, succinct format. Typically, this was completed through a set of reports or through verbal communication from section heads. With the development of the pilot, the senior managers could "drill down" and do some of their own follow-up, maximizing time utilization and reducing clutter since it was a user-driven system. Explanations, detailed data, maps, charts, documentation and ad hoc queries were all available online. All users viewed the same data with the opportunity to develop their own conclusions before contacting other staff to discuss the current conditions. Figure 4.4 shows an example of a pilot application.

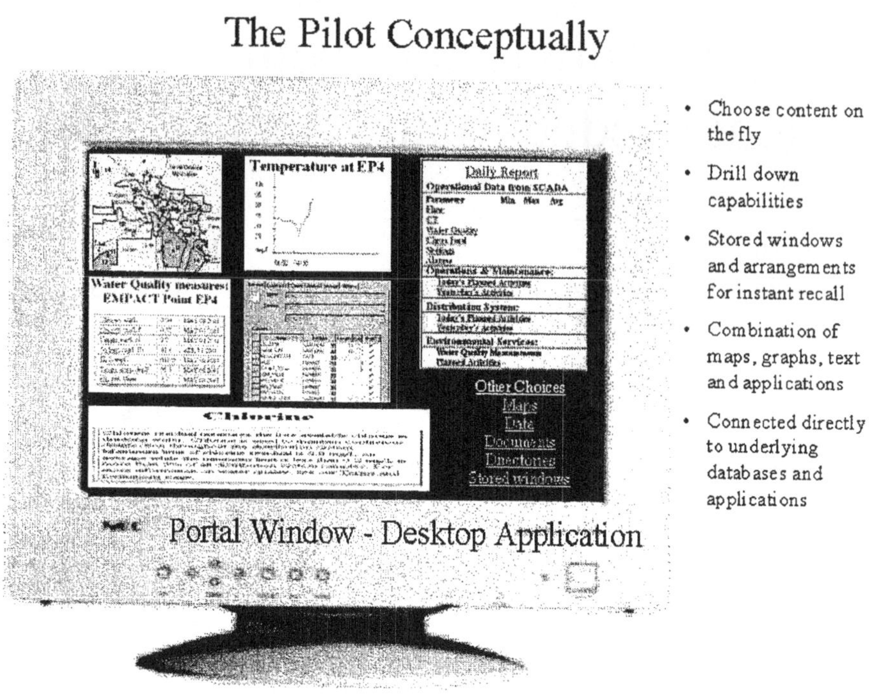

Figure 4.4 Example of a pilot application

In summary, the pilot system was intended to facilitate the problems of the morning brief. It included the following key attributes:

- Use of a data warehouse to manage the water quality related data
- Customizable set of water quality information automatically presented whenever the "desktop" was started
- The ability to "drill down" into information to get to the details of the raw data
- Ability to set triggers to automatically notify the manager of events
- Access to an automatically generated indexing of documents (records) to improve retrieval

IS KM FOR UTILITIES?

It is clear that Knowledge Management tools can address many of the barriers related to water quality issues and facilitate solutions for drinking water utilities. Knowledge Management can be implemented at many levels. In summary, Knowledge Management can:

- Facilitate the integration of different data sets
- Organize interfaces both for data input and reporting
- Force a cultural shift to computer centric information processing and access
- Facilitate standard Quality Assurance and control procedures
- Organize and disseminate raw data, data aggregates, information, and knowledge
- Facilitate communication among core systems
- Force the development of standards or reduce the need for standards by mapping disparate systems to one another
- Serve as a directory of available information
- Overcome information overload by organization and powerful search and discovery tools
- Integrate complex tools into a cohesive structure
- Serve as a powerful tool for standard reporting for compliance, internal communications, and customer relationships
- Expand the number and types of online data users

Knowledge Management benefits to an organization includebetter communications with customers and increased satisfaction, improved workflow and planning, cost reduction, product leadership, and operational excellence. Utilities will benefit from better customer communication, including increasing the number of calls handled per day, and increasing the number of calls resolved in the first "sitting" also will benefit from overall operational excellence including increasing productivity and yields at reduced labor rates. Utilities can mobilize important information and data for their staff and management to improve efficiency in operations, planning, decision support, and communications. The results will be streamlined operations, cost savings, and improved competitiveness.

However, there are many components to consider prior to the decision to implement a KMS. First, it cannot replace the development of good fundamental standards, documentation, and organization. Any KMS needs underlying data, standards, and organization.

Second, utility size should be considered before implementing a KMS. Full implementation of Knowledge Management may not be appropriate for small utilities serving less than 50,000 customers, but probably is appropriate for medium to large utilities. The costs may outweigh the benefits for smaller utilities as implementation of integrated information systems and Knowledge Management tools is an expensive proposition and requires significant investments in staff time. In small utilities, sophisticated tools probably will not be required to improve the access to and use of information. A well-designed information system with an adequate directory of resources will enable a few key people to act as information experts and help train users in how to access and use information. For infrequently used data, a single point of contact can act as the analyst and conduit of results required by others. While personal exchange of data works adequately in small utilities, as utility size increases, the effectiveness of personal data transfer decreases. The need for a more sophisticated system tends to be greater the larger the utility, thus making Knowledge Management tools more desirable for organizing, informing, and disseminating information.

Third, implementation of integrated information systems and Knowledge Management tools is an expensive proposition and requires significant investment in staff time. However, an integrated system, such as a KMS will reduce significantly the redundancy of information retained at a utility. This will result in a single point of contact, reduce transcription errors, and enable a few essential people to act as information experts.

Fourth, a comprehensive requirements analysis is necessary to define the information requirements, modes of interaction, and obstacles to both data access and transfer of pertinent data throughout utilities. The system should be able to support the goals and questions of the utility management. Through a process, a utility should hold focus group meetings and a joint requirements planning session with representatives of all stakeholders. Each utility can be organized different, so the utility should define the focus groups that must convene, emphasizing generators of water quality data, especially SCADA systems, field monitoring, and laboratory analyses. Utilities should document current practices for data management and processing, as well as planned improvements. The onsite analysis of a utility requires several important steps. Once the key stakeholders who represent the functional areas, e.g. source water treatment or maintenance, described previously have been identified, the utility should take the following steps.

1. Distribute questionnaires to the stakeholders to develop a good understanding of the information needs of the functional areas. This also helps prepare the participants for the focus group meetings.
2. Convene focus group meetings with each functional area individually to develop an in-depth understanding of the area's requirements and current practices. Document current practices for data management and processing, along with planned improvements. Explore and document barriers to effective data access.
3. Summarize the findings from the utility and return it to the participants for review and comment.

Finally, it requires a significant change in a utility's organizational culture to be successful. This cultural change often is more difficult than the investment and the challenge of gathering and organizing all the required information. The perception that data belong to one part of the organization, to the exclusion of others, is not acceptable and senior management must aggressively discourage it. The organization must make a commitment to become information-based. System implementation must include extensive training in the specific application of the system, especially for middle and upper managers. The utility must adopt standard operating procedures for data entry, processing, and documentation.

WHAT TOOLS ARE NECESSARY FOR KNOWLEDGE MANAGEMENT AT UTILITIES?

Knowledge Management tools such as data mining and data warehouses facilitate the organization and integration of an array of data and information into a form that can generate knowledge and support exploratory analysis. These methods and tools can provide information to the entire organization that was accessible only in individual departments. This data availability and cross-referencing can result in significant improvements in management decisions, planning, operations, and interactions with stakeholders. Currently, the utilities focus on the development of the core data sources needed to fuel Knowledge Management, including SCADA, LIMS, maintenance management, enterprise-wide databases, Customer Resource Management, GIS, and a suite of administrative systems. The project team recommends a core suite of software tools for medium to large utilities, as a basic functional system for water quality-related data and information. Once this basic infrastructure is in place, utilities will be able to move to the next level of aggregating all the data sources into knowledge. The construct of warehouses to perform data mining, portals for customized views of the information, and event mechanisms to automatically monitor the state of the utility will not be achievable without the core tools in place.

FUTURE RESEARCH

Future research needs may include:

- Development of industry-wide standards
- Development of integrated water quality databases
- Web-based interfaces for integrated information systems
- Improved processing, integration and dissemination of SCADA data

Compliance reporting currently forces organizations to report their data in a standard form, often in paper form. As utilities develop systems based on an industry-wide standard, more of the regulatory and compliance information process can be better automated. The ability to instantly analyze a national perspective then becomes a possibility.

APPENDIX A
KNOWLEDGE MANAGEMENT
LITERATURE REVIEW
CONTENTS

PAGE

INTRODUCTION

BACKGROUND

Drinking water utilities are complex business organizations that generate a significant amount of information of varying types including business, operations, water quality, planning and communications with stakeholders. Recently, national drinking water regulations require more data collection and reporting (U.S. EPA 1998). The amount of data and information collected along with the added requirements of federal and state regulations is resulting in an increased interest in the utilization of information in the management of drinking water utilities.

Historically, the main focus of water supply reliability has been assuring the quantity of available water supplies (Cromwell et al. 1997). The majority of the information collected was of limited use and application. In recent years, another water supply reliability goal has been added, assuring the quality of available water supplies (Cromwell et al. 1997). With new requirements and a perception that the data collected is a resource that can improve efficiencies and water quality, data access has taken on a new urgency. In order to meet these increased standards of efficiency and customer service, water suppliers are finding it necessary to redefine their business approaches. Management, analysis and dissemination of data and information are becoming strategic approaches to improve efficiencies and support daily decisions by operators and managers. The management and dissemination of the data are a major challenge.

Knowledge Management and Business Intelligence are processes and tools that allow businesses to increase workflow efficiency while reducing employee workload and labor hours. These processes and tools help employees manage, access and use information as a resource in all aspects of their work. This project evaluates the state and usefulness of these new two fields and their application to the drinking water industry.

OBJECTIVES

The objective of this task is to provide a report that reviews and details the current state of Knowledge Management and Business Intelligence and a summary of how these methodologies are currently being used in the drinking water industry. This document evaluates information management and knowledge-based technology for the capture, management and dissemination of data and information for drinking water utilities. A summary describing the capabilities and future paths for Knowledge Management is included. A bibliography is also presented regarding the most recent technologies in Knowledge Management and Business Intelligence.

REPORT ORGANIZATION

The second section provides a discussion of Knowledge Management and Business Intelligence processes and applications including data warehousing, on-line analytical processing (OLAP), data mining, portals, and records/document management. Operational systems including Supervisory Control and Data Acquisition (SCADA), Geographical Information Systems (GIS), and Laboratory Information Management Systems (LIMS), Enterprise Resource Planning (ERP) are also presented. The way these systems/tools affect customer support, planning, and operations is also discussed. Section 3 presents applications of these Knowledge Management tools to the drinking water, wastewater and power industries. Section 4 provides a summary of Knowledge Management tools.

KNOWLEDGE MANAGEMENT AND BUSINESS INTELLIGENCE REVIEW

Drinking water utilities share many characteristics of large commercial businesses and large government agencies. They have tasks that are universal, including billing, human resources, maintenance, customer support, operations and engineering. Unlike many commercial businesses they have major compliance requirements that involve careful monitoring and data management. As these requirements continue to expand and as the drinking water business becomes more competitive, proper use and leverage of available information will improve a utility's competitiveness, compliance reporting and water quality.

In an effort to assist drinking water utilities understand the status of information technologies that can support management, access and use of an array of information distributed throughout a utility, this project reviews the definition of the technologies. In addition it will explore case studies of utilities that are applying these technologies. Since few drinking water utilities are using these strategies and tools the search of literature has been extended into other utilities (e.g., electric and gas).

The drinking water industry typically has no shortage of data. On the contrary, so much data is available, it is difficult to manage, disseminate and apply. Usually the available data are distributed among an assortment of databases and documents. Along with the problems aggregating information within a utility, it is difficult to organize available information and maximize its usefulness in management operations and compliance reporting.

This section presents a review of Knowledge Management and Business Intelligence and the application of these tools and strategies. When Knowledge Management tools are properly applied, existing information is organized and mobilized, resulting in an increased utility of the information. The tools mobilize heretofore existing but inaccessible resources to support decisions, processes and information exchange.

WHAT IS KNOWLEDGE MANAGEMENT?

In order to define Knowledge Management, the collection, creation and organization of data, information, and knowledge, as well as and the overlap between them should be defined. Data is defined here as a single number or an alphanumeric string related to a measurement or observation. Examples of data are names, dates or values captured from observations. Data are raw values and do not provide a means of interpretation. When many data are combined together, they result in a data set or database. Data can be organized in either a structured or unstructured format. Data that are organized and stored in fields and records are called structured data, which may include names, addresses, and account numbers. Structured data may be stored in files, spreadsheets or databases. Data that are not organized into rows and columns or data objects are referred to as unstructured data. These may include word processing documents, computer reports, scanned documents, videos and e-mails.

Information is organized raw data that has been put into a context. It has meaning and delivers a message sometimes in a structured format. Information is often developed from analysis and interpretation of data. Both soft networks (e.g., a memorandum) and hard networks (e-mail, files, and wires) deliver information. Information generally has an influence on the person to whom it is transferred.

Knowledge takes raw materials (data and/or information) to the next level and can be viewed as a finished product. Knowledge can be contextual information contained within documents and archives, databases, policies and procedures, personal experiences, and expert insight. Thurman (1999) documents that in order to transform information into knowledge, it must be comparable to similar information, have known consequences, a connection to other items, and conversation (i.e., what other people think about this information).

The overwhelming amount of information, coupled with varied historical data management, results in a need for investments and technology to help achieve proper Knowledge Management. The outcome of Knowledge Management can create a greater competitive edge in the marketplace, improved service, improved quality and increased revenue.

Our research indicates a broad range of how Knowledge Management is perceived by different people. American Productivity and Quality Center (1996) and Firestone (1998b) view Knowledge Management as strategies and processes that are capable of identifying, capturing and maintaining an organization's knowledge. SAS Institute defines Knowledge Management as a system-based approach that is capable of capturing and leveraging institutional memory. Butler (1999) defines Knowledge Management as an integrated approach comprising of processes, technologies, attitudes, and systems which promotes and provides for adaptation and survival of the organization in a rapidly changing environmental workplace. Although discrepancies as to how this new field is faced, the main goal of Knowledge Management is to identify, capture, manage, analyze, and extract knowledge from various sources within an organization and place it into a structured, organized format (which may need to be refined over time) that can be shared.

It is important to distinguish between information management and Knowledge Management. While the definition of Knowledge Management is not clear, a distinction between Knowledge Management and information management is quite clear. Information management is the organization of data and information into a logical structure that facilitates the capture, quality assurance, management, analysis and dissemination of the data and information. It is the development of a structure and rules for organizing the data and information. Knowledge Management on the other hand is the development of relationships between the different data structures. It includes integration of many databases into a higher level understanding of how the data and information relate to one another. While Information management involves the organization and application of data and information to decision making, Knowledge Management relates the results of the analyses to one another developing insights and wisdom about the relationships of the results. Knowledge Management captures the insights of individuals and automates these insights into the processing of information and into observing and extracting relationships between different data in underlying information systems

Using the information provided by Business Intelligence (discussed in the next section) and associated technologies, Knowledge Management encourages the organization into the next necessary phase. This phase allows for effective marketing, increased customer service and management, and helps improve overall workflow within an organization. This can afford water utilities the goal of gaining a competitive edge, achieving operational excellence, and desired customer satisfaction.

WHAT IS BUSINESS INTELLIGENCE?

Business Intelligence is a set of methods and processes that improve business decisions using information from multiple sources and applying experience and assumptions to develop an accurate understanding of business dynamics (Butler 1999, Pilot Software 1999b). It is the gathering, management and analysis of data to produce information that is distributed to people throughout the organization to improve strategic and tactical decisions. Business Intelligence is facilitated by analytical and collaborative tools and applications including data warehousing, on-line analytical process (OLAP) and data mining.

Basic operational systems are systems utilized to automate routine tasks and are characterized by large volume of data. These systems are designed to "get data in" and create records of a specific routine without further processing. SCADA systems, utilized by the majority of water utilities, is one example of these systems. SCADA systems continuously record water pressure in the distribution systems that are transferred by telemetry to a database without any processing. On the other hand, Business Intelligence systems move beyond the basics of operational systems collecting and then storing data in a large repositories (e.g., data warehouse) with capabilities for exploration, analysis, and presentation of information. Due to the processing of the data, these systems are said to "get information out".

APPLICATION OF THE SUPPORT TECHNOLOGIES TO ASSIST WITH KNOWLEDGE MANAGEMENT AND BUSINESS INTELLIGENCE

The support technologies that assist with Knowledge Management and Business Intelligence include data warehousing, records and document management, data mining, OLAP, portals, enterprise resource planning, and customer relationship management.

Data Warehousing

Historically, businesses were unable or unwilling to place their data in archives since disk space was expensive. In recent years, while disk space has become more affordable, companies have realized that storing their data for analysis can lead to meaningful information for decision-makers within the company to try and keep a competitive edge (SAS Institute 1999c). To minimize the impact of these Business Intelligence inquiries (e.g., data mining and OLAP) on the performance of operational systems (e.g., billing systems, customer databases), Business Intelligence systems separate online transaction processing (e.g., "getting information in") from online analytical processing (e.g. "getting information out") by creating a separate repository of information. A data warehouse is this repository of data and information, that is isolated and optimized for data identification, access and analysis. It receives input from databases that are normally optimized for data capture and transaction processing. Data warehouse products organize both data and metadata.

The main function of a data warehouse is to consolidate data from various sources (e.g., customer databases, billing systems, etc.) into an environment optimized for analytical and information processing. The goal of a data warehouse is to successfully organize and manage data so that it can meet the mission to "get information out" (SAS Institute 1999b). Typically the information needed for decision support includes databases, maps, reports, memos and almost any other electronic form of information. For example, a data warehouse might be designed for decision support at a utility level combining together daily operational data from a variety of sources including dosages, flow rates and maintenance. Alternately, the warehouse could be designed for managing information relating to many utilities such as the consumption of energy versus water flow to commercial customers.

It is important to stress that the information organized in a warehouse goes well beyond simple joining of tables in relational databases. Data warehousing involves many disparate data sources, formats and levels of aggregation. A well designed warehouse organizes and controls the entire information process within an organization through data access and management. The warehouse facilitates the organization, processing and utilization of data and information through tools including OLAP, data mining, and query and reporting tools. A data warehouse requires continuous maintenance and updates to maximize its utility.

Approaches to Creating the Data Warehouse Environment

In general, there are two approaches recommended for the creation of a data warehouse (Firestone 1998a, SAS Institute 1997, Informix 1998). One approach is a central or enterprise data warehouse, which uses data directly from data archive systems and other sources, transferring data to departmental data warehouses (also known as data marts) or directly to users; that is, terminals that can be accessed by users. A second approach focuses solely on independent data marts importing data from any source. Data marts are smaller data warehouses that can be developed using the same software as an enterprise data warehouse, but focusing on the needs of an individual department.

In general, the creation of subject-oriented data marts are generally easier to manage as they usually contain smaller volumes of data than a central data warehouse and have fewer interrelationships. However, there are some disadvantages to implementing data marts. They may result in disparate reports since they are independent structures containing dissimilar data. In addition, extraction processes may not be as cost effective and place additional resource demands on operational systems.

However, the enterprise data warehouse approach can sometimes consume more time, money, and resources resulting from the large quantities of data and the complexity of mixing data and information from a variety of sources into a single system. In addition, since the scope of an enterprise warehouse can be broad, there may be more demands on the analysis portions of a system working to extract data, causing extraction problems.

To circumvent the problems associated with both enterprise data warehouses and data marts, an integrated data warehouse approach is recommended by SAS Institute as a solution to reduce operational costs and increase efficiency (SAS Institute 1998b).

Populating a Data Warehouse

Populating a data warehouse should be performed on a scheduled basis determined by the operational needs of the organization. Populating and updating a data warehouse begins with a data management phase including accessing and extracting data from all primary sources (those sources used to originally capture and organize the information destined for the warehouse). This step is followed by data organization including any necessary transformations and/or manipulations before the data is finally loaded into the warehouse (SAS Institute 1999c). The final step is data exploitation. Exploitation tools can help identify relationships of interest including OLAP multidimensional data viewers, data mining tools, or query and reporting tools (e.g., SQL query generators) (See Section 2.3.2).

Metadata is technical information that describes elements in the application including descriptors, tables, and columns (Davis and Joyner 1999). To develop a useful data warehouse, both physical metadata and informational metadata are required to support the investigation and analytical processes. Physical metadata defines the way an application stores or transforms data, such as the number or rows and columns within a table or any type of transformations applied to those rows or columns. Informational metadata describes the content or purpose of a table or the name of the responsible party for the information in the table.

Sources of Data for a Data Warehouse

While sources of data are numerous, including, but not limited to, e-mails, blueprints, correspondence, consultants, contract laboratories, regulating agencies, product suppliers and equipment manufacturers, for purposes of this report we have reviewed some of the larger sources of operational data. This includes operational systems such as independent billing systems for utilities, Supervisory Control and Data Acquisition Systems (SCADA), Laboratory Information

Systems (LIMS), Geographical Information Systems (GIS), on-line transactional processing (OLTP), enterprise resource planning (ERP), and customer relationship management (CRM) databases. Figure A.1 presents these operational environments and displays how they relate to Knowledge Management.

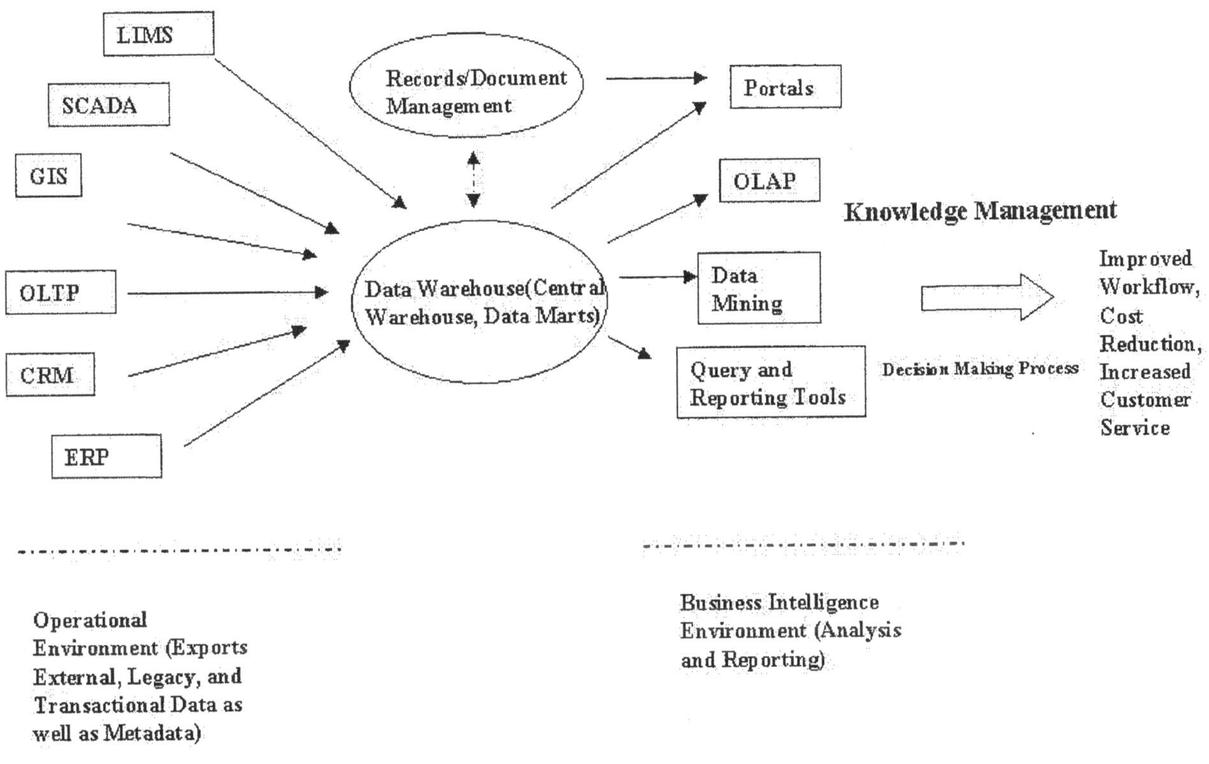

Figure A.1. Flow Chart of Knowledge Management

Supervisory Control and Data Acquisition Systems

One of the most critical and important aspects of the Operational Management Systems (OMS) for a water utility is transferring time series data automatically from field instrumentation via telemetry, data logger or other (e.g., fiber optic cables) systems to operational databases (Barnicle 1997, Casey et al. 1993). Currently, one of the most efficient methods for retrieving and integrating field data in a central system is through the use of SCADA systems. SCADA systems are used for the monitoring and control of operating networks in water systems, and for the management of wastewater processing facilities. Reasons for the use of this type of system include monitoring and controlling of pumping stations, water storage reservoirs and flow regulation valves. In general, SCADA is required by management for the following reasons:

- Reduce power costs
- Reduce staffing
- Reduce future capital requirements
- Improve level of service
- Avoid environmental incidents
- Comply with regulators requirements
- Obtain a competitive edge

Laboratory Information Management Systems

Laboratory Information Management Systems (LIMS) have been used in commercial laboratories for many years, although these computer systems have somewhat different uses in a water utility laboratory (Paraira and Tejero 1999). LIMS are used for water quality testing, analysis, and reporting. The primary purpose of a LIMS system is to provide water quality measurements, and allow for management capabilities to facilitate the mission of the water utility. Another main function of these systems is their ability to serve as a repository of water quality information, allowing users to observe trends in data. Data is stored and retrieved such that laboratory staff can track the progress of a given batch of samples while data users can receive answers to water quality inquiries (Hertz et al. 1998).

Geographic Information Systems

A Geographical Information System (GIS) is a computer system capable of assembling, storing, manipulating, and displaying geographically referenced information (e.g., data identified according to their locations). Overall, GIS technology can be used for scientific environmental investigations, water resource management, watershed monitoring, and development planning (Ward 1996). Some examples of the specific uses of GIS include:

- Calculation of emergency response times in the event of a natural disaster by emergency planners.
- Assistance in locating wetlands that need protection from pollution.
- Location of a new water well in a watershed after hydrologic tabular data is converted to map-like form.
- Production of images including maps, drawings, animations, and other cartographic products.
- Production of detailed diagrams or maps of main hydraulic networks, etc.

Enterprise Resource Planning

Enterprise Resource Planning (ERP) systems are software systems that enable organizations to better manage financial and human resources and allow the integration of daily operational functions and tasks (SAS Institute 1999d). Currently, human resource departments, financial institutions and manufacturing are three of the main businesses that use ERP. ERP systems are being implemented at organizations with a goal to facilitate and automate processes, increase productivity, improve customer service, and reduce operating costs.

Since ERP systems are useful for process-management purposes, it can be advantageous for managers in charge of daily operations to utilize them since they enable departments (e.g. human resources, payroll, and financial systems) to relate to one another and share data and knowledge. In a drinking water utility, ERP could be used to maintain employee records, training schedules, required training, scheduled maintenance, work load allocations, scheduling tasks and assigning personnel to complete those tasks.

Customer Relationship Management

Customer Relationship Management (CRM) is a broad term encompassing a variety of strategies used by companies to maintain customer loyalty, organizational profitability and increased sales (SAS Institute 1999b). To remain competitive, companies should continuously learn from interactions with each individual customer and be prepared to enthusiastically respond to information and knowledge gained from those interactions. The best CRM systems include: comprehensive account management functions, contact profile and history, automated quotes and correspondence, forecast, order and contract generation, instant access to historical account summary, and Internet communication (Pilot Software 1999b).

To achieve a balanced CRM approach, there are three primary management technologies: operational, analytical, and collaborative (Tekeitz and McCarthy 1999). Operational CRM is the focus of many organizational initiatives. It deals with front-office customer issues including sales, marketing, and customer service through the use of integration between front office and back office, interconnected systems. Analytical CRM is the analysis of data from the operational CRM that facilitates collaborative CRM. Analytical CRM is directly related to data warehouse architecture during analytical applications (e.g., data mining, OLAP). Collaborative CRM facilitates interactions between customers and organizations. It is the application of shared services (e.g., e-mail, conferencing) attempting to improve customer service and customer retention by communication and organization.

This group of software organizes an important but difficult group of knowledge into a useable format that can be combined with other critical data to improve the management of the utility and the perception of clients about the utility's ability to serve them. Examples of this are when a customer relations person can easily pull up laboratory results that were generated to address a customer complaint or concern. This will be difficult without a good CRM attached to a Knowledge Management system and data warehouse.

Another supporting tool, Computer Output to Laser Disk (COLD) can also be used to handle customer comments in a searchable repository/archive allowing these statements to be displayed and printed immediately in response to customer inquiries (Tower Technology 1998).

TOOLS TO EXTRACT AND REFORMAT OUTPUT FROM A DATA WAREHOUSE

Since a data warehouse is optimized for the extraction and use of information through analysis, the tools used to extract, reformat and analyze the data and information play a pivotal role in the success of the warehouse. This section presents a brief summary of the analytical tools used to extract data from a warehouse that are unique to Knowledge Management. These tools include on-line analytical processing (OLAP), data mining, portals, and query and reporting tools. Relationships not apparent from the raw data can be explored through these predictive tools and exploratory data mining techniques (e.g., regression analysis, decision trees, clustering, sequencing). These tools provide products that assist management with the decision making process (Figure A.1).

Most of the Knowledge Management query and reporting tools discussed in this section shield organizations from Structured Query Language (SQL) syntax and any potential problems with the complex construction of an SQL query. However, since the creation of queries through the use of SQL also enables the user to search for patterns or trends in data, SQL may be warranted in some cases and it should not be overlooked.

On-line Analytical Processing (OLAP)

Once the data warehouse is built, Online Analytical Processing (OLAP) provides a methodology for supporting automatic storage, reformatting and processing of the data going into the warehouse. OLAP is an analytical information processing technology, generally integrated to the data warehouse, with the ability to examine data from multiple dimensions and is performed "on-line" and interactively (SAS Institute 1996b). Through the use of transformations and calculations upon existing data, OLAP creates business information from existing data. This information can either be stored as part of the warehouse or presented to the user for consideration and determination of the fate of the new information.

Although OLAP tools present data to users in a multidimensional fashion, the databases that support this technology can be relational, multidimensional, and vertical. Relational databases are organized horizontally in rows, which are called "records" (Pilot Software 1999a, Firestone 1997). Each record contains related information organized into "fields". Vertical databases are based on data that is stored vertically in columns. Multidimensional databases group fields into dimensions, and then dimensions into multidimensional arrays. Each format and approach facilitates a different way of viewing data and understanding its content. In addition each format provides insight and acts as precursors to various analytical tools. For example, a relational format is appropriate for a sophisticated statistical package. Multidimensional data would be a useful output for a management report or for a data analyst exploring relationships in the treatment process. OLAP can also be used to develop graphics to evaluate the data as they are processed, without specifying the perspective or level of detail in advance. For example, a drinking water utility may want to consider a few dimensions consisting of water quality analyte levels both spatially (e.g. over the regions of the distribution system) and temporally (over time).

One perspective of the data might be a snapshot of disinfection levels in the distribution system while another view could be the variation of disinfection levels in the process train. In addition to dimensions, this technology allows us to investigate information at varying levels of detail. Analysis starts by examining the big picture and as it progresses, allows for modifications and the exact level of detail to be explored later on in a more detailed sense.

In summary, OLAP can be seen as a hardware/software configuration that works as an observation and query tool to uncover new relationships/information from the existing data. It also acts as a support tool for analysis and confirmation at the review level (SAS Institute 1996b).

Data Mining

While data warehouses store and organize data, there is still a need to efficiently filter through vast warehouses of data. Data mining is the main process by which trends can be identified and evaluated, which turns data into information, and provides visual representations for data exploration, statistical analysis and modeling, and an assessment of the results (SAS Institute 1998a). For example, given large quantities of data, often it is not practical to analyze the database as a whole. Instead, a subset of the data (or sample) is used, allowing for quicker analysis and modeling information. Data mining then uses that data to answers questions, which frequently lead to more specific questions through an iterative process. This type of process leads to a competitive edge for most businesses (Fayad and Uthurusamy 1996).

SAS Institute (1998a) divides data mining into five stages referred to as SEMMA (Sample, Explore, Modify, Model, and Assess) to provide steps in which the data mining process can operate.

These general guidelines facilitate exploratory statistical and visualization techniques for business analysts. In addition, they assist with the selection and transformation of the most significant predictive variables, and model the variables to predict outcomes while confirming the accuracy of the model. The SEMMA methodology includes:

- Sample the data by creating data sets large enough to contain the important information, yet small enough for quick processing.
- Explore the data to gain a better understanding and more ideas by trying to identify relationships, trends, and anomalies.
- Modify the data (e.g., transforming) variables for the model selection process.
- Model the data, which may predict a certain outcome.
- Assess the data through the use of the results of the data mining process.

When considering data mining, it is important to keep in mind that even if powerful data mining tools are implemented, relationships within warehouses may not always be evident (SAS Institute 1998a). This may happen if the data are not properly aggregated or prepared for analysis. This can also occur if relationships within the data are too complex to be seen by observation alone, and if the databases are too large to be processed economically as a whole. However, these issues can also be resolved. For example, to solve problems associated with data aggregations, data warehouses can be subdivided into data marts.

In general, data mining should be approached as a process, as opposed to a series of analyses, and it should provide a flexible methodology and user-friendly multi-level (e.g., descriptive, advanced) statistical analyses and reporting techniques (SAS Institute 1999a).

Data mining can be an effective tool when problems emerge and a utility must try to determine their causes. It is also useful when operators observe conditions that occur rarely but repeatedly. If they want to understand how the conditions were handled in past instances, analysis of the data can help determine approaches that were used and the effectiveness of those strategies. Often this type of an evaluation cannot be done because raw data are not available and the required analytical tools are not integrated with the data. Knowledge Management tools applied to data mining would allow for multiple occurrences to be identified, and the results of the operations strategies to be combined with the data describing the conditions to reach conclusions and recommend option for current conditions. This approach emphasizes that retiring expertise does not mean loss of corporate memory.

Portals

Modern Enterprise Information Portals (EIPs) allow an organization to customize the view of an underlying information system or warehouse (Firestone 1999a). Each group or perspective within a corporation or a wider user community can be customized to deliver the subset of information required in a format and level of content detail, appropriate to a particular group or individual. EIPs provide total access (essentially a single gateway) to multiple forms of critical data as a starting point for a user (e.g. customer, employee, etc). Some types of portals even support e-mail and scheduling functions. It is important to note that most EIPs require a dedicated server for accessing data (Firestone 1999b). They also usually depend on a host of underlying technologies (e.g., databases, transactional systems, OLAP, query and reporting tools, data mining, search engines, and web browsers) which can make implementation more difficult.

Portals can used effectively to act as the window into a warehouse. They can also be used to automate Knowledge Management strategies that are customized to individual users while accessing a single warehouse or a limited number of data marts. Portal applications include electronic document retrieval for customer self-service, business documents (e.g. contracts, policies, forms, bills, invoices), public document access, and web-posting. In addition, portal applications may also include e-mail and content capture, warehouse and folder management, as well as correspondence and case management.

Records Management / Document Management

Records and documents are among the most valuable assets within an organization. Both are part of the warehouse and in some cases they record and track things in the warehouse or created through the use of Knowledge Management tools. Records can be any information essential to a company and may reside in numerous locations, including on-site and off-site locations and document archives. Records need to be stored and retrieved (Treev 1998). Often a record or a document has a history, which needs to be preserved. Records can include (but are not limited to) databases, records in a database, e-mail, documents, images, graphs, database records, data files, spreadsheets, scanned documents, etc. Essentially anything that can be stored in an electronic file is considered a record. Documents are usually analog records that are stored either as hard copy or as scanned images. These images can be both electronic or analog (e.g., microfiche). Documents can include items like letters, permits, applications, maps, faxes, PDF files, and microfilm.

Some documents and records have a useful life span that should not be exceeded. Records and document management help determine when and how a document should be destroyed. Records management focuses on any type of record. Records management focuses on electronic or analog media (paper, pictures, photos, blueprints) while document management focuses on the analog records. The document management system itself can be either an electronic or a paper system. Document management applications manage (e.g. creation, capture, indexing, and retrieval), store and distribute documents throughout their lifecycle. It also adds version control, shared workflow, and structured communications.

Historically, records management and document management were separate entities. However, over the past five years, these applications have been merging. Document management can usually be handled in records management systems. However, records cannot always be handled in document management systems.

Document management and imaging both assist with document management and move documents from individual desktop computers to large central repositories. Integrated document management (e.g., separate engines) may be necessary since documents are captured, stored and indexed in different ways. For instance, some documents are captured and stored as images (or discrete pages) whereas others are captured from outside sources and stored as large reports. In addition, some are generated in office applications and stored as files. Document imaging converts paper documents into electronic formats by organizing, scanning, indexing and storing optical images. Both of these applications facilitate the location and retrieval of images and other documents.

SUMMARY

In summary, combining an operational environment with a Business Intelligence environment through the use of a data warehouse and associated tools allow businesses to gain a competitive advantage. Some benefits may include increased customer satisfaction, improved workflow and planning, and cost reduction. The strategies and tools discussed above facilitate the organization and integration of an array of data and information into a form that can be used to generate knowledge and support exploratory analysis. These methods, tools and strategies provide information to the entire organization that normally was accessible only in individual departments. This availability and cross referencing of data can result in significant improvements in management decisions, planning, operations and interactions with stakeholders. These improvements are not only contingent on the installation of new hardware and software, but also depend on a change in culture and use of data and information. The installation of these tools in the absence of a change in the business will not have a significant effect, and in some cases may result in poorer performance.

APPLICATIONS OF KNOWLEDGE MANAGEMENT AND BUSINESS INTELLIGENCE TOOLS TO THE DRINKING WATER, WASTEWATER AND POWER INDUSTRY

This section focuses on documented case studies of the application of Knowledge Management and Business Intelligence to utilities. The original intent was to focus on drinking water utilities. However, due to the few references located the authors have expanded their search into associated utilities to find sufficient case studies to contrast approaches and reach general conclusions. Case studies describing types of successful systems for utilities are documented. In addition, a matrix of relevant software, associated platforms, and software capabilities is presented. Both components should allow for a better understanding of how these technologies might be used in the drinking water industry.

CASE STUDIES

Each case study includes varying levels of information including the challenge the organization was facing, the old approach to the problem, what solution they implemented, what functionality the system has, what were the costs and benefits of the approach, and whether the implementation was successful. The case studies are presented by the major project undertaken by the utility.

Data Warehousing

Columbus Water Works

A Columbus-based water company, Columbus Water Works serves safe drinking water and treats wastewater for almost 200,000 residents in the Columbus, Georgia region. A data warehouse was implemented to act as a central source of all information from all of their applications. This "star schema" database allowed for a reduction in the number of necessary data transfers and also created a main dataset for reporting purposes. The term "star schema" is a general database term, but applies more to Data Warehousing as opposed to transactional databases. A star schema is a highly denormalized, query-centric model. The star schema is built for simplicity and speed. The assumption behind it is that the database is static with no updates being performed online, which is the norm for data warehouses. A star schema allows for each stars (e.g., sector of management) data to be independent but also allows for integration of star data through the central data warehouse. The Columbus Water Works warehouse has six primary "stars" including customers, employees, inventory, measurements, work orders, and accounting. Columbus documents that the main benefit of the management of data through the warehouse for a utility is cost control. The data warehouse has allowed them to collect and distribute information between applications, better manage expenses, monitor data trends over time, and explore any unexpected variations.

Tennessee Valley Authority

A southern utility, Tennessee Valley Authority (TVA), serves more than 3.4 million customers across Tennessee, Alabama, Mississippi, Kentucky, Virginia, North Carolina, and Georgia. TVA instituted a successful data warehousing system, providing employees with greater accessibility to "real-time" data, and allowing for continuous data log updates (e.g., water routing logs) to the backup control center. While the system has been quite successful, they document training as one factor to seriously consider when implementing a data warehouse in a utility. They suggest the development of guidelines for program design, documentation and maintenance.

U.S. Geological Survey (USGS)

The National Water Quality Assessment Program (NAWQA), created in 1991 and sponsored since by USGS, collects physiochemical and biological data from various regions in the country. The data collected is stored in a database currently housing 6.5 million records. The database, Oracle9iAS, is one of the components of the USGS warehouse. USGS staff and external customers use associated database applications including querying and data retrieval. The data search is easily handled by the system through web browsing, although certain types of queries are not as quick as needed. The main goal of the retrieved data is to support analyses, which are critical to assess water quality in many different American regions.

Records & Document Management And Imaging

Essex and Suffolk Water Implements Advanced Customer Service System

A major U.K. utility, Essex & Suffolk Water that supplies drinking water to approximately 1.75 million people living in Essex, south Norfolk and north Suffolk is implementing an advanced customer relationship management system. They manage roughly 650,000 homes and 46,000 businesses. They are implementing RIVERS (Retrieving Information Via Electronic Routing Systems), to improve the utility's service levels both to its commercial and residential customers. RIVERS can provide quicker response to customer inquiries as well as an anticipated increase in workflow productivity.

The decision to implement RIVERS was based on the inefficiency of the existing manual systems as they were very paper intensive with a microfilm archive system that slowed down the speed of responses in the call centers. An electronic system enables improved customer service and more effective resource management.

This system will automate many aspects of customer correspondence handling by scanning incoming correspondence, indexing and then automatically routing to the appropriate customer teams. This can allow for same day responses on the day inquiries are received. This system allows for immediate electronic access to all the information needed to resolve issues, including a customer's complete case history and will improve the quality of replies to customer calls.

Essex & Suffolk Water representatives have documented that the system is currently operating in the Customer Service department and is anticipated they will institute the system in their Customer Accounts department as well. They also document that their staff has accepted this new electronic technology very well.

Yorkshire Water

Another United Kingdom utility, Yorkshire Water is responsible for serving more than 4.5 million customers within the Yorkshire region and has a staff of 3,000 that processes approximately two million phone calls and 1.5 million postal items per year. They wanted to improve service and reduce costs, a feat that would have been unattainable based on the previous existing paper-based system and a microfilm archive system.

The utility automated its correspondence handling by initiating the WAVES project (Work Allocation Viewing and Execution System). A Tower high performance imaging and case management (workflow) system was installed to replace the existing microfilm document management system. Through the use of scanning and indexing correspondences for the call center agents, issues can be quickly resolved as the Customer Service representatives have immediate electronic access to each customer's case history ensuring higher quality service with fewer staff.

In summary, the implementation of an imaging and case management (workflow) system reduced operating costs and increased the quality and timeliness of customer service. It also changed the working environment by allowing staff to work from the scanned image instead of the paper original.

South East Water Limited

A southeastern Melbourne, Australia company, South East Water Limited, provides water and sewer services to 500,000 properties in the area. As a result of losing or misplacing paper records and some 100 year old paper records crumbling, South East Water decided to make some changes. Information regarding a property owner's sewer connections to Melbourne's main sewer lines is vital so customers won't hit a sewer pipe while working on their own property.

In an effort to better manage these records, they implemented Wang's OPEN/Image software to store, retrieve, and manipulate their records. This system has been quite successful in increasing customer service by decreasing the call times from 10 minutes to one minute. In addition, when new changes are made to properties, they are scanned into the system so they will be available for future use. South East Water also reported substantial cost savings in both storage space and labor.

Ontario Hydro

A Canadian electricity company, Ontario Hydro is responsible for providing access to the 16,000 Material Safety Data Sheets (MSDSs) on the hazardous materials used at the utility to all employees. These sheets originated at the 2,100 companies that produced the hazardous chemicals and were then sent to Ontario Hydro. This process involved periodic updates as well. Ontario Hydro had many problems with employees losing the MSDSs or spilling oil on them while on the floor, causing concern that employees did not have the most accurate information.

They implemented Spicer's Imagenation imaging software and linked it to their existing database allowing for images to be scanned or faxed which facilitates viewing the MSDSs for the employees. The system also allows for cover sheets for all images indicating the accession number and date to ensure higher quality, current information. All images are maintained on CD and in an effort to remain current, every few months the CDs are updated and sent to the 300 work sites for company use.

Omaha Public Power District

A southeast Nebraska utility, Omaha Public Power District (OPPD) provides electricity to 270,000 homes and businesses in the area. OPPD wanted to improve customer service and reduce operating costs, so they decided to replace the Records Management Department's Computer Output to Microfiche (COM) system. They implemented three OnBase processing modules: OnBase COLD, OnBase Document Imaging, and OnBase CD Authoring. On a daily basis, OnBase COLD is able to process fifty different types of mainframe reports including over 18,000 billing statements. Used in conjunction with a scanner, OnBase Imaging scans and indexes field service technician paper forms.

The system changes were quite successful in improving customer service by allowing customer service representatives direct access to customer account information at their own workstations. It also helped reduce operating expenses as they no longer had to pay a service bureau over $100,000 a year to write daily mainframe records to microfiche since OnBase CD Authoring saves backup copies of all stored documents to CD.

Baltimore Gas and Electric

A Maryland utility, Baltimore Gas and Electric, responsible for providing gas and electric service to 1.6 million customers, implemented the File Magic Plus document management software suite. This decision was made since damage control crews had to constantly request drawings indicating the location of underground gas mains whenever they were sent to a site that had a damaged underground utility (e.g., leaks). The process of retrieving drawings formerly involved processing requests, photocopying, and mailing the drawings to other sites. This process sometimes took a few days to complete.

File Magic Plus allows the documents to be scanned into the system with notations on the drawings. It also provides workers with instant access to drawings from their workstations. In addition, documents can be shared across the network. Retrieval times have decreased substantially and an improvement in overall workflow efficiency has been documented.

Integral Energy

Integral Energy, responsible for delivering power to 1.3 million customers, has 20 offices with 1,500 employees connected by a WAN. Integral Energy had an overwhelming amount of paper work, mostly in the form of faxes, which sometimes hindered the efficiency of the customer service process. In an effort to increase the efficiency of the entire network and to improve customer service, Integral redirected their incoming faxes to the Tower Fax Gateway, which delivered the faxes to the appropriate recipient. They also used Tower workflows to facilitate their daily step-by-step processing.

PSE&G

The New Jersey based utility Public Service Electric and Gas (PSE&G) is a highly regulated nuclear power plant that has to maintain a high level of documentation. It is critical that all documentation including engineering drawings, procedural documents, instruction and vendor manuals remain current. Since the previous system for maintaining, updating, and distributing new drawings was a lengthy process, PSE&G decided to implement the ViewStar system.

ViewStar is capable of storing and handling large images and allows PSE&G to scan revised documents and replace old documents, maintaining one version of each document. This allows cost and labor savings, as they do not need to make copies, maintain separate files, or sign distribution slips. They are also considering a text search component from Verity that will allow searching text documents and images for information. PSE&G document that they will see a return on their investment in three to four years.

OTAY Water District

A San Diego water and sewer service agency, OTAY Water District, provides water and/or sewer needs of close to 100,000 people residing in a 129-mile square area of San Diego County. This includes southern El Cajon, La Mesa, Rancho San Diego, Jamul, Spring Valley, Bonita, eastern Chula Vista and Otay Mesa along the Mexico border. They chose a new information management system, Myriad viewing software, allowing employees direct access to maps on their laptop, eliminating the need for the GIS department to reply to requests for drawings from employees in different departments. In addition to allowing users to view parcel drawings, the viewer has other powerful features and allows access to associated information (e.g., bitmaps, AutoCAD files, WordPerfect documents, Excel spreadsheets, and other scanned drawings). This change has increased overall workflow efficiency.

Furnas Electric

A Brazilian company, Furnas Electric, was looking to improve the work instructions on the factory floor. Furnas implemented Myriad viewing software to streamline their internal processes, automate special orders, and reduce scheduling problems. Myriad allows for assembly steps to be displayed graphically allowing operators to know exactly what is involved in building a specific piece. Myriad also reduced the need to have different drawings on file, eliminating the need for permanent storage of assembly drawings. This system also allows employees to change drawings as they work. It is also capable of layering individual steps to create assembly instructions and also for the creation of customized products on screen.

Gulf Power Company

A Pensacola, Florida electric utility, the Gulf Power Company supplies electricity to over 300,000 customers in 71 towns and communities throughout northwest Florida. Gulf Power wanted to improve customer service, time-savings, and increase the safety of their operations. Since field workers needed to know the location of underground power lines, Gulf Power equipped their trucks with Myriad viewing software for instant access. When there are unexpected power outages, the field workers can have access by a lighted screen, eliminating the need for searching for paper maps, calling the control room, even going back to the office to obtain another photocopy of a map. This ultimately allowed for better customer service. Historically the average amount of time a customer was without power was 98 minutes per year. Now it has been reduced to 38.6 minutes per year.

Workflow / Remote Access

Sydney Water

Sydney Water, responsible for the provision and maintenance of Sydney's water and wastewater facilities, manages over 1.5 million Sewerage Service Diagrams (SSDs). SSDs are used when maintaining or modifying water and sewerage services by providing service, sewerage, boundary, and environmental information and are necessary additions to a contract for the sale of a property. Previously obtaining a SSD was a lengthy process since it involved searching for a property in an online database, retrieving the appropriate book and copying the page. In addition, since access to SSDs were limited at each office, processing requests could require a half hour to several hours worth of effort.

Sydney Water installed seven Tower Systems to improve customer service by reducing the delays in providing copies of SSDs and create a more cost-effective operational system. The Tower systems decreased search times and allow maintenance crews easy remote access to the entire catalog of SSD documents. Through the use of laptop computers, crews can view and select SSDs, which can then be faxed to emergency vehicles, allowing them to locate problems (e.g., sewerage service breakage, chemical spill) within minutes. In addition, the process of editing SSDs has also improved by integrating the Tower System with faster editing software, Auto-trol. In less than ten minutes, documents can be edited on screen and electronically re-filed compared to the months it used to take to manually edit the diagram.

In addition, within minutes Sydney Water is able to transmit large-scale engineering drawings to regular fax machines through a process referred to as tiling. Many sheets will be printed that can be placed together for an exact size replication of the original drawing. As a result of the implementation of this image/document management system, timeliness and quality of customer service in all areas has improved.

General Electric

General Electric, a world leader in electronics, communications, electrical and mechanical engineering, stores a wide variety of documents (e.g. invoices, delivery dockets, product information), images, engineering drawings, and computer generated files for the GE Industrial Systems.

Remote access, including retrieval and scanning, is provided from the central data archive site via Tower's remote cache server technology and direct WAN connections to workstations at the larger and smaller sites respectively. In addition, they use Enterprise Report Management to directly upload documents from the mainframe into the Tower system as they are processed for customers. If invoices need to be retrieved, an overlay of the information is made over a blank invoice thus eliminating the step of scanning the invoice after it is sent.

Customers are also able to request documents including invoices and delivery dockets through an automated system via an interface between GE's Interactive Voice Response technology and the Tower system.

Customer Service / Billing

Wessex Water Services Ltd.

A European water utility, Wessex Water Services Ltd. supporting 1.1. million water customers and treating sewage for 2.5 million customers, hired Arbor Energy to customize a billing engine with debt collection capabilities. This type of system allows for a customer self-care option via the Internet as well as improved effectiveness of the customer service representatives. The Arbor/BP Product and Arbor/OM products provide for multi-service billing, as well as increased customer service and an integrated workflow product, respectively. In addition, they implemented Strategist, an analysis application designed to enhance customer knowledge. Throughout theses changes, Wessex Water has been recognized as one of the most efficient water and wastewater companies in the United Kingdom providing outstanding customer service through its newly implemented systems.

Three Valleys Water

A British water company, Three Valleys Water treats water to 3 million people living in the region of North London. The treatment capacity of the plant averages 800 million liters of water per day. Three Valleys Water is implementing a Customer Relationship Management (CRM) system designed by Tower Technology. The CRM system is based on document and process automation. The goals of this system are to support staff operations with online access to all information needed to handle customer issues and pending cases, catalog and process service requests, and process invoices and statements. Overall, the company expects that the new CRM system will improve quality and responsiveness of the service provided to customers.

Michigan Consolidated Gas Company

Michigan Consolidated Gas Company distributes natural gas to approximately 1.2 million residential, commercial and industrial customers. IBM developed and implemented a system, Interactive Voice Response (IVR), which integrated the utility's voice and data systems. The IVR provides information to customers including balance and payment histories, budget, billing, enrollment, payment agreements, service requests and repair status. This automated service increases productivity in customer relations allowing the personnel more time to handle other issues.

Other (GIS and Database Related Studies)

Georgia Power Company, GPU Energy, Niagara Mohawk Power Corporation, Northeastern States Power Company, and Pennsylvania Power and Light Company

A suite of companies employed Coherent Networks to implement SMARTMAPS software to facilitate workflow and improve customer service by creating real-time links between the utility, its customers, its workforce and its network. An Atlanta based power company, Georgia Power, a part of the largest producer of electricity in the U.S. – the Southern Company, is responsible for providing electricity to 1.7 million customers within the state of Georgia is a good example of this. SMARTMAPS allowed for a direct link between Georgia Power's customer database and AUTO/Cad outside plant facility files. Both Niagara Mohawk Power Corporation and GPU Energy also used SMARTMAPS to convert pole, secondary, primary, and underground maps as well as electric distribution map originals to a GIS environment.

In addition, Northern States Power Company customized the SMARTMAPS software to allow for conversion of electric and gas distribution maps. It also enabled the combination of facility information with government and commercial landbase data. Pacific Power and Light Company used SMARTMAPS to convert numerous maps, records and archive data files.

Pinellas County Utilities

Pinellas County's (and surrounding areas) water system, sewer system, reclaimed water system, solid waste operations, and recycling systems are maintained by Pinellas County Utilities (PCU). PCU was interested in implementing applications that might help streamline businesses so they initiated an Information System Analysis (ISA). From this, they implemented an ORACLE database structure that eliminated the use of file versions and one that does not call for the user to know where tables are actually located. This enables protection of data integrity as well as efficient data management. Through the implementation of a graphics package (SHL version) that utilizes the Oracle database, 300,000 pieces of utility infrastructure and 500,000 land parcels are now referenced through the use of GIS. In summary, the addition of a GIS package allowed for better infrastructure management, map publishing and improved business decision making.

SUMMARY

Each case study found in the literature included varying levels of information including the challenge that organization was facing, the old approach to the problem, and what solution they implemented. In addition, the functionality of the system, the costs and benefits of the approach and whether the implementation was successful was also documented. Table A.1 lists utilities with success stories using Knowledge Management tools.

Table A.1
Review of Case Studies Found in the Literature

Utility	Knowledge Management Tool
Columbus Water Works	Data Warehousing
Tennessee Valley Authority	Data Warehousing
U.S. Geological Survey	Data Warehousing
Essex and Suffolk Water	Records & Document Management and Imaging
Yorkshire Water	Records & Document Management and Imaging
South East Water Limited	Records & Document Management and Imaging
Ontario Hydro	Records & Document Management and Imaging
Omaha Public Power District	Records & Document Management and Imaging
Baltimore Gas and Electric	Records & Document Management and Imaging
Integral Energy	Records & Document Management and Imaging
Public Service Electric and Gas	Records & Document Management and Imaging
OTAY Water District	Records & Document Management and Imaging
Furnas Electric	Records & Document Management and Imaging
Gulf Power Company	Records & Document Management and Imaging
Sydney Water	Workflow/Remote Access
General Electric	Workflow/Remote Access
Wessex Water Services Ltd.	Customer Service/Billing
Three Valleys Water	Customer Service/Billing
Michigan Consolidated Gas Company	Customer Service/Billing
Georgia Power Company	Other (GIS)
GPU Energy	Other (GIS)
Niagara Mohawk Power Corporation	Other (GIS)
Northeastern States Power Company	Other (GIS)
Pennsylvania Power and Light Company	Other (GIS)
Pinellas County Utilities	Other (ORACLE and GIS)

In general, the literature case studies demonstrated that utilities are working toward increased work efficiency and cost reduction while trying to increase customer support. This has been achieved primarily through the use of data warehouse, record/document management and imaging, and GIS. While portals, OLAP, and data mining are very important Knowledge Management tools, the use of these more traditional knowledge management tools was not observed in these case studies.

Software Technology to Implement Knowledge Management Tools

Since it is generally difficult for any single vendor to provide all of the necessary information technology solutions, it is essential to recognize the application and software packages that compliment each other, providing such a solution. Table A.2 presents a matrix approach evaluating the various technology tools, which can be utilized to accomplish the process of Knowledge Management. In addition, their capabilities and necessary platforms are indicated.

A "Knowledge Management system" (KMS) is not a package that can be purchased like a statistical package or accounting package. A KMS is something developed and applied as a solution for a particular organization. It consists of many components selected for the particular capabilities that can be integrated to meet particular requirements. Prior to the fairly recent popularity and sophistication of Enterprise Resource Planning systems such as SAP, resource management was a similar custom application or a collection of tools integrated in a custom solution.

Using the matrix in Table A.2, an organization might begin their process of developing a KMS by selecting the functions needed to improve efficiency of the organization. Regardless of choices, there will be compromises faced and prioritized. The one common core is the need for data housed in a database, which could be one or more data warehouses or perhaps direct access to a set of databases. Data warehouses are primarily a performance solution. Additional capabilities draw upon this data using the tools described, such as data mining, to help determine trends, portals to assist in selective views of the information, agents to monitor for events, and records management to assist in find and tracking information stored "somewhere" in the system.

SUMMARY AND PHASE 1 RECOMMENDATIONS

Organizations are moving towards a Knowledge Management environment to address both specific and general needs to improve the utility of information. The specific stimuli include but are not limited to: lost knowledge from the downsizing and restructuring of organizations, increased employee-customer contact at all levels, competition for customer relationships, globalization, and the need for increased workflow efficiency (APQC, 1996, APQC 2000).

Through research of numerous organizations, APQC (2000) identified six major strategies for Knowledge Management including:

- *Knowledge Management as a business strategy* - useful for companies to compete and grow
- *Transfer of knowledge and best practices* - useful for organizations that want to improve operations, products, or services
- *Customer-focused knowledge* – focuses on knowing the customers needs and preferences, while improving communication and service
- *Personal responsibility for knowledge* – used by companies who realize that their employees are their most valuable asset
- *Intellectual asset management* – emphasizes intellectual assets including patents, technologies, operational and management practices in order to gain a competitive edge
- *Innovation and knowledge creation* – emphasizes basic applied research and development

Knowledge Management benefits to an organization include an improved customer relationship, product leadership and operational excellence. Utilities would benefit significantly from the customer-focused knowledge and KM as a business strategy approach, including but not limited to, increasing the number of calls handled per day and the number of calls resolved in the first "sitting". Utilities would also benefit from overall operational excellence including increasing productivity and yields at reduced labor rates and increased production cycle time. Utilities could mobilize important information and data for their staff and management to improve efficiencies in operations, planning, decision support and communications. The results would be streamlined operations, cost savings and improved competitiveness.

It is clear from this project that the types of information tools that need to be considered in prioritizing Knowledge Management systems are available. However, there is general consensus that no single system can provide a Knowledge Management system. Designing such a system will require careful thought and re-organization of workflow.

First, the right software/hardware approach needs to be implemented. Since it is generally difficult for any single vendor to provide all of the necessary information technology solutions, it is essential to recognize the applications and software packages that compliment each other and provide such a solution. This report provides a matrix approach evaluating the various technology tools, which can be utilized to accomplish the process of Knowledge Management.

Next, designating roles within the organization and focusing on organizational structure including identifying specific roles are essential to facilitate the generation, sharing, and management of knowledge. Roles should include people who create knowledge, those in the Knowledge Management infrastructure and those who actually apply the knowledge (APQC 2000). The creators of knowledge would include external sources, publishers, analysts, and network members. For example, typical data warehouse analyst roles could include a project manager, a warehouse architect, a warehouse administrator, and a data administrator. Roles within the KM infrastructure include publishing coordinators, knowledge organizers and architects, information technology, and perhaps help desk attendants. The final roles of those who apply knowledge include teams who are led by full-time knowledge managers.

Business Intelligence, when applied along with a shift in corporate culture, can have significant positive effects on the operation of a utility. It is critical to understand that the transition to a knowledge-based strategy is an expensive undertaking that needs to include proper hardware, software, personnel, policies and corporate commitment. Instituting Knowledge Management systems needs to begin with a clear understanding of requirements and priorities. The actual implementation is initially costly. Some returns on investment will occur very quickly in improved efficiencies and the ability to produce compliance reports and interactions with regulators. Large term improvements in efficiencies and reduction in costs will take longer but can be easily predicted and demonstrated. Some of these improvements will include staff reductions and dramatic improvements in efficiencies. Areas likely to first be affected through increases in efficiency include maintenance, operation and customer support. Historically challenging tasks like tracking paper documents can be easily addressed and resolved using Business Intelligence and Knowledge Management strategies.

ANNOTATED BIBLIOGRAPHY

American Productivity & Quality Center (APQC). 1997a. Competitive and Business Intelligence: Leveraging Information for Action.

This report discusses the use of competitive intelligence (CI) by organizations and enterprises for strategic planning and key decision making. The report is a result of a benchmark study involving a variety of enterprises that were researched to provide an in-depth look at the importance, critical success factors, and best practices in competitive intelligence. Organizational examples and graphs support seven major findings listed in the report.

American Productivity & Quality Center (APQC). 1997b. Using Information Technology to Support Knowledge Management.

This report contains the result of an APQC consortium study involving twenty-six organizations detailing how enterprises have applied information technology to Knowledge Management. Strategies, approaches, and tools used by leading Knowledge Management practitioners are presented and discussed in detail.

American Productivity & Quality Center (APQC). 1998. Expanding Knowledge Management Externally: Putting Your Knowledge to Work for Customers.

This work reports the results of a benchmarking study conducted by APQC on how innovative companies successfully share knowledge with customers. The main focuses are on areas such as the customer-focused knowledge strategy, knowledge content, measurement, and technology. The 15 findings are explored in detail throughout the report.

American Productivity & Quality Center (APQC). 2000a. Knowledge Management.

This guide includes a thorough understanding of the components of a successful Knowledge Management program. It provides mechanisms to identify and examine current status, understand the components of a successful Knowledge Management initiative, and determine implementation strategies.

American Productivity & Quality Center (APQC). 2000b. Successfully Implementing Knowledge Management.
This report, based on a Knowledge Management (KM) consortium benchmarking study by APQC, discusses how early KM adopters implemented their Knowledge Management initiatives. These case studies also show how these enterprises identify any performance gaps and consequent improvements. Findings provide enterprises with a better understanding of issues and challenges in implementing KM initiatives.

Bacheldor, B. 1999. Portals Make Business Sense. InformationWeek on Line.
The emphasis of this article is on the interest of companies about portals and their contents. Portals are reported as tools that can give employees, business partners, and customers a central access point to critical data, including Web sites, internal reports, news feeds even enterprise applications such as ERP and CRM software.

Barnicle, A. 1997. Implementing Monitoring, Control and Management Information Systems in a Major Water Utility. Water Supply, 15:2, pp. 75-99.
The paper discusses the benefits of using a standard methodology for designing and implementing SCADA systems at Severn Trent Water, England (water and wastewater treatment). In addition, it discusses the use and benefits of telemetry data as the basis for an operational and corporate management information system.

Butler, S. 1999. Knowledge Management Directions: From BI to Knowledge Management. Data Management (DM) Review.
This on-line paper draws a parallel between Business Intelligence and Knowledge Management. The author cites the most known applications and tools used in both fields.

Casey, R.J. and Grinney, S.A. 1993. An Operational Information Database Based on the Automated Processing of Time Series Available From Telemetry Control Systems. Water Science and Technology, 28:11/12, pp. 111-121.
One of the most important aspects of this paper focuses on the development and subsequent

management of a central database of operational information based largely on the automated processing of time series data retrieved from the Thames Water Utilities Limited.

Castillo, E., S.K. Keefe, R.S. Raucher, and S.J. Rubin. 1997. Feasibility of Small System Restructuring to Facilitate SDWA Compliance. Published by the AWWA Research Foundation and American Water Works Association. ISBN 0-89867-900-1.
The purpose of this work is to evaluate the potential for small water systems to physically interconnect or engage in cooperative ventures with larger water systems. The General Accounting Office suggested this restructuring in 1994 as the best solution for small water systems that are not in compliance with the Safe Drinking Water Act Amendments of 1986. This study did not include case studies. Data used in analyses were obtained from the Federal Reporting Data System (FRDS).

Collins, A. G. 1992. Information Processing Coupled with Expert Systems for Water Treatment Plants. ISA Transactions, 31:1, pp. 61-72.
This paper focuses on the use of predictive models and diagnostic experts to enhance the traditional feedback control systems used in water treatment plants. These predictive models may be either regression relationships or neural networks.

Cromwell, J., D. Owen, J. Dyksen, and E. Means. 1997. Managerial Assessment of Water Quality and System Reliability. Published by the AWWA Research Foundation and American Water Works Association. ISBN 0-89867-905-2.
This report provides an in depth examination of the main issues in the drinking water industry since it is facing new reliability challenges to assure the quality of water to be supplied to customers. Views of forty utility managers were collected and used to produce a consensus statement on water quality and system reliability.

Curtice, D., L. Jentgen, and R. Ward. 1997. Energy and Water Quality Management System. Published by the AWWA Research Foundation and American Water Works Association. SBN 0-9648877-2-X.
This report focuses on identifying the benefits of water and electric utilities participating in a Forum

to find solutions to their information systems problems as well as identifying which characteristics make a Forum successful. This is accomplished by several workshops at one participating utility, East Bay Municipal Utility District (EBMUD).

Daniel, P.A. 1998. Balancing Multiple Water Quality Objectives. Published by the AWWA Research Foundation and American Water Works Association. ISBN 0-89867-940-0.
This report focuses on the identification and management of multiple and interdependent water quality objectives related to treatment process optimization. A thorough evaluation and prioritization of these water quality objectives was completed using two case studies for the San Francisco Water Department and the Santa Clara Water District.

Davis, J. and Joyner, E. 1999. Successful Customer Relationship Management: Why ERP, Data Warehousing, Decision Support, and Metadata Matter. A SAS Institute White Paper.
This paper states that customer relationship management would be successful depending on the quality of data and data-management processes. To correctly manage those relationships, the company must employ specific decision support systems such as enterprise resource planning (ERP) and data warehousing. The authors also recognize that data and metadata associated with these various systems are central to an organization's future.

DELPHI CONSULTING GROUP. 1998. How Tower Meets the Need for KM. Delphi Consulting Group White Paper.
This paper outlines Tower's approach to managing the knowledge of an organization. The Tower approach is to pre-integrate component technologies into basic industry applications such as documents, images, and COLD reports.

EMA Services, Inc. 1997. The Utility Business Architecture: Designing for Change. Published by the AWWA Research Foundation and American Water Works Association. ISBN 0-89867-903-6.
One of the main goals of this report is to discuss how and why Utility Business Architecture (UBA) methods and tools can be used to develop an integrated model to be utilized by water utilities to analyze, redefine, and improve their business changes and operations performance. The UBA is

based on a concept called object technology. These objects (i.e. water utility business elements), can be represented by a variety of functions including a pump or a customer's bill, or even conceptual process, for example, dealing with an emergency call. Various workshops with 10 large and midsize public and private water utility provided requirements, input, and data to develop and test the model. The final utility business model consists of functions relevant to products or services for the customer. These functions include water production, treatment, transmission, and distribution.

Fayyad, U. and Uthurusamy, R. 1996. Data Mining and Knowledge Discovery in Databases. Communications of the ACM, 39:11, pp. 24-26.
This article discusses the importance of data mining and knowledge discovery (e.g., management technologies for data analysis). The objectives and challenges of this field are briefly presented and explored.

Firestone, J. M. 1997. Evaluating OLAP Alternatives. A White Paper from Executive Information Systems, Inc.
This paper presents an insightful analysis and identification of OLAP tools including Multidimensional (MDOLAP), relational (ROLAP), and the Sybase Vertical (VTOLAP) for data warehousing. This paper also provides a set of criteria for product evaluation in specific project contexts.

Firestone, J. M. 1998a. Architectural Evolution in Data Warehousing and Distributed Knowledge Management Architecture. A White Paper from Executive Information Systems, Inc.
This paper describes DSS/data warehouse system architectural evolution in response to the growing complexity of the enterprise DSS environment.

Firestone, J.M. 1998b. META Group and the Future of Knowledge Management. A White Paper from Executive Information Systems, Inc.
The author describes the relevance of Knowledge Management as a distinct organizational business process. He also lists an extensive set of requirements for defining KM applications.

Firestone, J. M. 1999a. Enterprise Information Portals and Enterprise Knowledge Portals. A White Paper from Executive Information Systems, Inc.

The author makes a parallel between information and knowledge portals. As stated by the author, the business world is not interested in generating more information. It is not information that they are interested in, but knowledge. Knowledge provides increased return on investment (ROI), increased effectiveness, and a sound basis for decision making.

Firestone, J. M. 1999b. Defining the Enterprise Information Portal. A White Paper from Executive Information Systems, Inc.

This white paper is an excerpt from a report entitled "Approaching Enterprise Information Portals". It summarizes Enterprise Information Portals (EIP) definitions by a variety of professionals and experts.

Gardner, S. 1998. Building the Data Warehouse. Communications of the ACM, 41:9:1998, pp. 52-60.

This white paper provides a summary of questions that project managers should ask their companies executives and themselves in building a data warehouse. It also provides guidelines needed to sort out the answers.

Hertz, C., Bardsley, A., Kurisko, P. and Senft, M. 1998. System Helps Agency Manage Laboratory Information. Waterworld, October.

This paper is a case study describing the use of Laboratory Information Management System (LIMS) by the central laboratory of Philadelphia Suburban Water Company. The LIMS, from LabVantage Solutions, serves a repository of water quality information and handles hundred of sample locations from more than 20 water systems.

INFORMIX. 1998. An Integrated Approach to Enterprise Resource Planning and Data Warehouse Systems. An INFORMIX White Paper.
This white paper discusses the general trends within ERP systems and why a sound data warehousing strategy which integrates with operational ERP systems, is becoming essential to managing a successful company.

Inmon, W.H. 1996. The Data Warehouse and Data Mining. Communications of the ACM, 39:11, pp. 49-50.
This paper provides a brief discussion on the relationship between data warehouse and data mining emphasizing that a data warehouse greatly improves the chances of success in data mining.

Kingdom, B., J. Knapp, P. LaChance, and M. Olstein. 1996. Performance Benchmarking for Water Utilities. Published by the AWWA Research Foundation and American Water Works Association. ISBN 0-89867-877-3.
This report examines the use of performance benchmarking to be applied in the water industry.

Meserve, J. 1999. Preparing your Firm for Corporate Portals. NetworkWorldFusion.
This article provides a brief overview of corporate portals and some of its applications. A number of case studies are provided to show the importance of using portals as a fast and efficient way of accessing, handling and managing corporate documents and data.

O'Dell C., Grayson, J.C, and Essaides, N. 1998. American Productivity & Quality Center (APQC). If Only we Knew what we Know: The Transfer of Internal Knowledge and Best Practice.
This book provides an insightful analysis of Knowledge Management practices applied by many different companies. It provides case studies with concrete examples on how knowledge is transferred within an organization.

O'Leary, D.E. 1998. Enterprise Knowledge Management. IEEE Computer Society, March, pp. 54-61.
A variety of factors are pushing enterprises to explore better methods for Knowledge Management.

The main purpose of enterprise KM is to make knowledge accessible and reusable to the enterprise. This article surveys some components of this field.

Paraira, M. and Tejero, J. 1999. How Water and Environmental Laboratories Benefit from LIMS Software Installation. International Environmental Technology, July/August.
This paper is a study case describing the benefits of using Laboratory Information Management System by Aguas de Barcelona, the central company of an international group, the AGBAR Group. The main activities of this company are water treatment, supply, sampling, and analysis.

PILOT SOFTWARE. 1999a. Discovering Hidden Value in Your Data Warehouse. 1999. A Pilot Software White Paper.
This paper provides an overview of data mining and its relevance to today's business environment. Topics included are foundation, scope, techniques, architecture for data mining and how it works. In addition, examples of applications are discussed to illustrate the use of this technology to extract hidden predictive information from large databases.

PILOT SOFTWARE. 1999b. An Introduction to OLAP Multidimensional Terminology and Technology. A PILOT Software White Paper.
This paper focuses on the use of OLAP technology for Business Intelligence applications. A parallel is drawn between relational and multidimensional databases emphasizing the importance of the latter to support OLAP for viewing and analyzing business data.

PILOT SOFTWARE. 1999c. Customer Intelligence: Business Intelligence for CRM Solutions. A Pilot Software White Paper.
This paper states that a well-planned CRM solution would benefit nearly every department in a company, principally sales, marketing and customer support. To support this statement, this paper provides information on data warehousing and Business Intelligence, customer intelligence solutions (scope and focus), and integrated analysis solutions.

PILOT SOFTWARE. 1999d. Web-Based Decision Support: Scaling to Support Very Large User Communities. A Pilot Software White Paper.
This paper exploits the importance of the relationship between decision support solutions and users. The technology employed has been optimized from attacking complex analytical problems for one or a few users to a Very Large User Communities (VLUC). For example, the more users (e.g., managers, within the organization) of such a solution, the more likely decisions made will lead to the desired result.

Prescott, D. R., W. Underwood, M. Kindl. 1995. Functional Baseline Requirements and Data Elements for Records Management Application Software. Army Research Laboratory, Software Technology Branch, Atlanta, GA.
This report specifies functional requirements and data elements for a Records Management Application (RMA). The proposed requirements and data elements can be used to support the management of information assets as records.

Riekert, W., Mayer-Foll, R., and Wiest, G. 1997. Management of Data and Services in the Environmental Information System (UIS) of Baden-Wurttemberg, SIGMOD Record, 26:1, pp. 22-26.
This paper is a case study describing the Environmental Information System of Baden-Wurttemberg, Germany. Distributed sources of information from various supplier sites are combined in an integrated network.

SAS INSTITUTE. 1996a. OLAP tools and Techniques within the SAS System. Technology to Match the Way People Think. A SAS Institute White Paper.
This white paper details the SAS OLAP solution and discusses how IT personnel can use SAS software to integrate online analytical processing into their existing data warehouse technologies.

SAS INSTITUTE. 1996b. A Formula for OLAP Success. Using On-Line Analytical Processing to Transform Data Into Knowledge. A SAS Institute White Paper.
This article is an informative overview of OLAP technologies for exploration, analysis, and

reporting of trends found in business data. It covers general aspects starting with Business Intelligence systems, OLAP, multidimensional analysis, and finally providing some examples for applications of this tool.

SAS INSTITUTE. 1997. Data Warehouse Administration. Produced in Association with the Data Warehouse Network. A SAS Institute White Paper.
The first part of this paper discusses requirements and procedures that are needed for administering data warehouses. The second part focuses on the SAS Institute's approach and products for managing the warehousing environment.

SAS INSTITUTE. 1998a. Data Mining and the Case for Sampling. Solving Business Problems using SAS Enterprise Miner Software. A SAS Institute Best Practices Paper.
This work discusses the relevancy of data mining and sampling as a statistically valid practice to investigate complex relationships that exist in large databases. The topics explored include data mining as part of the Business Intelligence Cycle statistical analyses using sampling techniques and utilization of sampling in data mining. Finally, sampling techniques are illustrated by presenting a case study where the company used one of the Institute products, SAS Enterprise Miner, to analyze their data.

SAS INSTITUTE. 1998b. SAS Rapid Warehousing Methodology. A SAS Institute White Paper.
The emphasis of this paper is on the SAS methodology for designing data warehouses. The methodology defines the responsibilities of resources assigned to the project and clearly sets expectations for each phase of the project.

SAS INSTITUTE. 1999a. Finding the Solution to Data Mining. A Map of the Features and Components of SAS Enterprise Miner Software Version 3. A SAS Institute White Paper.
This paper describes the data mining software tool developed by SAS Institute, Enterprise Miner. This tool uses a graphical user interface (GUI) allowing the user to follow all the steps in data mining starting with the sampling of data and ending with the dissemination of the resulting information.

SAS INSTITUTE. 1999b. Implementing a CRM-Based Campaign Management Strategy. A SAS Institute White Paper.

SAS Institute discusses the technological and business issues involved in implementing a marketing campaign management initiative using advanced data management and analysis techniques in managing customer relationships.

SAS INSTITUTE. 1999c. The Quality Data Warehouse. Serving the Analytical Needs of the Manufacturing. A SAS Institute White Paper.

This paper covers aspects and use of data warehouses for improving quality in industries. It also discusses steps for building a Quality Data Warehouse, type of information that is delivered with this data warehouse, and how that information is delivered.

SAS INSTITUTE. 1999d. Intelligent Warehousing for ERP Systems: Leveraging SAP AG's R/3. A SAS Institute White Paper.

This white paper examines the relationship between ERP and information delivery/decision support, i.e., how the SAS solution for Intelligent Warehousing enables organizations that have invested in the ERP system to get at their SAP AG R/3 data. Ways in which SAS solutions help these organizations to achieve significant business advantages are also discussed.

Tekeitz, F. and McCarthy, R.L. 1999. Analytical Customer Relationship Management. A White Paper from Sybase, Inc.

This paper examines a new approach to implementing a Business Intelligence solution focused on the analytical aspects of CRM, (e.g., analysis of data created by the operational side of CRM and legacy applications). Results of these analyses will help businesses to reach theirs customers more effectively.

TOWER TECHNOLOGY. 1998. Computer Output to Laser Disk: What is mean by the term "Output"? And does it Matter? A Tower Technology White Paper.

This paper discusses the types of customer statement data that should be stored on Computer Output to Laser Disk (COLD). COLD is as an integrated part of the imaging and document management system currently used by Tower Systems.

TREEV. 1998. TREEV: The New Paradigm in Document Management. A TREEV White Paper.
This paper presents the approach employed by TREEV for managing document and other unstructured data types. TREEV utilizes industrial imaging, COLD, workflow, and document management engines for rapid, cost-effective development of document management solutions.

Thurman, S. 1999. Knowledge Management Means Career Opportunity. Intercom. 46:7 p.19-21.
This paper initially explains the difference between data, information and knowledge. In addition, it details the relevancy of the Knowledge Management field for enterprises.

Tkach, D.S. 1998. Information Mining with the IBM Intelligent Miner Family. An IBM Software White Paper.
This white paper describes the text mining operations and techniques as they are implemented in the IBM Intelligent Miner Family, and emphasizes some of the applications of this technology. Text mining is the process where it is possible to extract unknown information from any source including transactions, e-mail, web pages, and others.

Torrence, D. and J.S. Perry. 1997. Exploiting Key Answers from your Data Warehouse using SAS Enterprise Reporter Software. A SAS Institute White Paper.
This paper examines a reporting tool designed by SAS Institute, SAS Enterprise Reporter Software, to be used for data preparation and administration. Data is extracted from the data warehouse and then analyzed before being reported. This tool allows quick performance in creating reports and avoiding the use of a programming language, which can sometimes be cost and time consuming.

VanWingen, R. 2000. Requirements for Electronic Records Management and their Significance.
This is a PowerPoint presentation from Tower Software's Records & Document Management Seminar on January 14, 2000. This presentation was retrieved from the Tower Software site (http://www.ustrim.com/trim/sem14012000.htm). It discusses topics such as requirements and

usefulness of Electronic Records Management. A comparison between Document and Records Management is also presented.

Ward, G.D. 1996. International Report: Network Management Systems. Water Supply, 14:3/4, pp. 121-155.
The author compiles information retrieved from National Reports on network management systems. These National Reports were received from fifteen countries. The major topics discussed are the area of development of hardware and software, asset information systems, operational planning systems, network modeling, monitoring and control, network management, integration of techniques, and change management.

Yeager, T. 1992. Rural Water/Wastewater Study Vol 1. Background, Terminology, and Recommendations. Rural Electric Research. Project 91-12.
This report provides the background on small community water and wastewater systems. This information can potentially be used by Rural Electric Cooperatives (RECs) that operate and manage these systems to ensure that they comply with the Safe Drinking Water Act (SDWA) and Clean Water Act (CWA). In addition, this report summarizes what can be accomplished by RECs to help small water and wastewater utilities with technical, financial, and administrative assistance in developing new projects and also on projects that have been completed.

Yeager, T. 1992. Rural Water/Wastewater Study Vol 2. Case Studies and Management Issues. Rural Electric Research. Project 91-12.
This companion volume presents an overview on groundwater and surface water resources, and point of entry/point of use (POE/POU) devices. It also provides a more detailed explanation of issues that are of particular interest to Rural Electric Cooperatives (RECs) as they begin aiding water and wastewater utilities. One of the issues discussed is the correlation between federal and state enforcement actions and the service territories of each REC. Other issues presented include the experiences that RECs have had in assisting water and wastewater utilities.

SELECTED URLS

Selected URLs on Data Warehousing, Data Mining and OLAP:

Data Warehouse Institute – http://www.dw-institute.com

Data Warehouse Institute, Lessons from Experts - http://www.dw-institute.com/lessons/index.htm

Data Warehousing Information Center – http://intelligententerprise.com/dwframe.shtml

Foundations of Data Warehouse Quality – http://www.dbnet.ece.ntua.gr/~dwq

IBM's Page – http://www-4.ibm.com/software/data/busn-intel

ORACLE's Page – http://www.oracle.com/datawarehouse/index.html

PILOT Software's Page - http://www.pilotsw.com/news/index.htm

SAS Institute's Page - http://www.sas.com/technologies/data_warehouse/index.html

http://www.sas.com/technologies/data_mining/index.html

http://www.sas.com/technologies/olap/index.html

Stanford Data Warehousing Publications - http://www-db.stanford.edu/warehousing/publications.html

Selected URLs on Knowledge Management and Business Intelligence:

American Productivity and Quality Center – http://www.apqc.org

Data Management (DM) Review – http://www.dmreview.com

IBM's Page on Business Intelligence – http://www-4.ibm.com/software/data/busn-intel

KM Forum – http://www.km-forum.org

Knowledge Transfer International – http://www.ktic.com

Knowledge Sharing – http://www-ksl.stanford.edu/knowledge-sharing/papers

ORACLE's Page - http://www.oracle.com/applications/bi/index.html?content.html

Summary of Resources – http://www.brint.com/orglrng.htm

SVEIBY's Page – http://www.sveiby.com.au

APPENDIX B
UTILITY QUESTIONNAIRE

Preliminary Information Assessment

Name:

Division:

Focus Group:
(For example: Information Services, Operations, Engineering, Maintenance, etc.)

All questions below relate to the capture processing, analysis and interpretation of data and information except where specifically noted otherwise. We are trying to get an initial understanding of how information is used by your utility.

If you do not know an answer, please indicate so and suggest how we can get this information. If the question is not applicable please indicate this on the form.

1. General

1.1. Describe the primary function of your work unit. This answer should describe what your division or department does for the utility. Include an explanation of your mission and how it fits in with the entire utility. This question refers to more than just information.	

1.2. Please describe the process you follow in your work.

How do you get the job done you described above? Include non-information-related tasks but focus on information capture and use.

For example: Our job is to control the operation of the treatment plants. We capture data from the SCADA system, visual observation and manual measurements. We also get input from the water resources department regarding projections of source water input quality. This input comes to us by a telephone call.

2. Infrastructure

In this section we are trying to understand what kinds of computer systems you have and what software you run on these computers. We are also trying to understand what kind of computer operation you have within the utility.

2.1. How many information specialists do you currently employ in your functional group?

These are professional information system people including hardware, software, communications and network specialists.

2.2. What professional Information Technology resources are available from the City, state or other organization? Are you required to use these services? Are you satisfied with these services? Why or why not? We are trying to understand if there is an umbrella organization that is responsible for providing Information technology services to your utility. For example: We are required to work on the municipal information system and network. We must use the Municipal databases and must adhere to their standards for hardware and software. They have a staff of 200 professional that provide exceptional service. They are always on time and deliver maintenance as needed.	
2.3. Do you have specific security requirements? If yes, please specify. Often Utilities and business in general have specific security requirements that are essential. For example: we need to have dial up or internet access to our SCADA system but we must make sure that no one outside the utility can view or control the SCADA from the outside.	
2.4. Do you have specific backup requirements? How are these done currently? What are the risks if these data are lost? Utilities have critical data that are needed for billing, operations and compliance reporting. Loss of these data could be catastrophic to the utility. For example: Our water quality results are generated once and then stored on a central database. These Data are only available in one place and are required for compliance reporting and operations. They are currently backed up by ____	

2.5. What kind of computer systems do you use (PC, PC Field notebooks/units, Macintosh, Unix, Mainframe, etc.)? Please describe all the computer systems that you use including computer attached to instruments. Describe your desktop computers and the central computer servers used by your utility	
2.6. What information systems do you currently maintain or use? Include databases, models, custom software, and desktop utilities. Current versions? Please list the software, models, databases etc. that you use. Indicate which are routinely used and which are used rarely. We are trying to understand the types of software you use and the level of experience of the users within you functional group. For example: We all use Microsoft Office on desktop machine that is running Windows 2000. The Network runs under Microsoft NT. We Use Outlook for e-mail and Access for simple database. We have two customized databases that run under MS ACCESS 1) Wells, 2) Water quality. The central servers run ORACLE (version 8). We have ORACLE databases for Customer Complaints, Maintenance Management and compliance reports, etc.	
2.7. Do you currently use Geographic Information System software (AutoCAD, ArcCAD, ArcView, ARC/INFO, MapInfo, etc.) to perform any of your functions? If so, what software and what functions? Describe any mapping or spatial analysis tools used by your utility.	

2.8. Are there other tools, programs, etc. that you would like to use, but are unable to do so? What is needed (availability, training, customization, etc.)?

In many instances people know they have needs but are unable to meet those needs for a variety of reasons. Please describe these. We are trying to understand opportunities in which information or functionality is currently not available but could or should be.

3. Data Generated by Others

In this section we want to establish other sources of data that are critical or useful to you in performing your job.

3.1. What types of data do you use which are generated or managed by other groups, either internal or external to the utility? Who are the other groups?

Specify the types of data.

For example: We use hydraulic data for our source water, weather data, planned communities from building permits etc. Please indicate the sources of the data that you indicated above

3.2. How do you access their data? How often do you access the data?

Please indicate how you get these data. Do you get them by dialing into a system, over the internet, on a CD, etc? Do you pay for the data or are they free?

For example: We access the weather data from the National Climactic Data Center over the internet. The data are free and we download the data quarterly.

3.3. Please describe the processing done with these data. After you obtain these data what analyses are done with the data. For example: After we down load the rainfall data for our watershed we take a weekly average which is used to ___	
3.4. Do you or anyone else in your organization store and maintain these data after you receive them from the source? Please indicate whether or not you store the raw data or the calculated results described above and describe the process for storing the data and who else uses the information. For example: After taking the weekly averages we put this information into our watershed database. The information is used for compliance reporting and in long-term watershed modeling.	
3.5. Are there any other data that you know of which would be useful to you but are not currently available (or not readily available)? Please describe these data and whom you think maintains these data. Please describe the barrier to accessing these data. What benefits would be achieved if these data were available?	

4. Data and information generated by your functional Group for your own use or for use by others.

The purpose of this section is to document the data and information that your functional group generates for the Utility. These data and information include results used both within and outside the utility.

4.1. List and briefly describe the major products or outcomes of your work. Focus on Information but also identify outcomes and products that are not specifically information. For example: We produce drinking water of high quality meeting or exceeding all regulatory and utility requirements. We also produce data that are used to evaluate the compliance with the regulations and reports summarizing the analysis of these data.	
4.2. List and describe any ancillary products that are useful to you or other stakeholders. For example: We also produce logs of equipment operations that are used by the maintenance personnel to perform planned maintenance. These equipment logs are also used by engineering in evaluating future upgrades and designs.	
4.3. What regulatory agencies receive reports based on your work (generated by either your or by other functional groups)? Who prepares the reports? Please describe the target agency for the report, the content and how the report is generated.	

4.4. List the key technical outputs (Reports, Graphs, Slides, Multimedia, Pamphlets, Videos) generated or needed by your work unit. What are the reports, graphs, figures, or electronic materials that your group produces? How are they produced? Where do you get the data? How are the products created? For example: We publish two reports each month. One is a summary of operation and water quality for the internal personnel. We get data from the central Water Quality database, from the SCADA and from hand written records. The information is all submitted in hard copy to a technical writer who puts the report together. A similar report is generate for our commercial customers but in less technical detail. It is created as a subset of the first report.	
4.5. Do you create maps to include in reports? Please describe and provide samples.	
4.6. What types of data are you responsible for disseminating to other work units or stakeholders? Please describe these products.	
4.7. Who are the other stakeholders who depend on you for data and information?	
4.8. How are the data disseminated to these other groups? If you are responsible for supplying data to the public or to customers please describe the data and the mechanism for delivering the data.	
4.9. Please describe the data that you currently maintain that you think would be useful to others that they are not using. What are the barriers to them using these data?	

4.10. Please estimate (gross estimates are adequate) the volumes of data generated and used by your functional group. Include units (e.g., megabytes/day, gigabytes/month, etc.)

This request is for a gross estimate. If you do not know the amount of data that you generate just enter that you don't know.

5. Locational Information

The objective of this section is to determine if there is a need for Geographic information Systems or the ability to map outputs.

5.1. Do you create/use databases that have fields containing location information such as address, township-range-section-quarters, parcel numbers, subdivision names, or block/lot numbers? If so, describe the databases and their source?

5.2. Do you use maps to perform your job function? If so, please list the function, the maps used, and their source.

How do you get maps? Are they created electronically or do you work with paper maps? If paper maps, where do you get them?

5.3. If you are not currently using maps, do you see a need for maps?

Please explain what maps would be useful.

6. SCADA System Description

The objective of this section is to obtain detailed information about your utility's SCADA system. This section should be filled out by someone knowledgeable about the data and information that results from the SCADA system.

6.1. What communications method is used to gather remote site data such as wells and elevated tank levels? (Telephone lines, radio, WAN etc.)	
6.2. How many remote sites are currently monitored?	
6.3. What instrumentation exists at the source of the raw water such as wells, lakes and etc.?	
6.4. What other Utilities or organizations are available to share communication links? (Wastewater, electric, etc.)	
6.5. Who is responsible for the maintenance of the SCADA system? Is the Information Services group utilized for any computer maintenance related to the SCADA system?	
6.6. Is the information collected by the SCADA system automatically generated into a regulatory agency report?	
6.7. Are there multiple plants that have the SCADA system on a LAN?	
6.8. What communication protocols are presently used in the plant SCADA system What PLC or DCS systems are utilized (Modbus, DF1, Tiway, etc.)	

6.9. How many Tag names are in the current SCADA system? (analog and digital inputs and outputs)	
6.10. How many personnel need to see the plant process data (other than reports) besides the shift plant operator? Please list these people and their job titles.	
6.11. Is the plant staffed in the control room for 24-hour operation?	
6.12. Is there a need for SCADA alarms to be monitored other than at the control room? (Pagers, radios, telephone).	
7. Any other comments or issues?	

Glossary

SCADA
Supervisory Control and Data Acquisition
A computer that is connected via a communication link that allows data exchange from a local control device such as a CPU, **PLC** or **RTU**. The control was traditional done at the **PLC** level (when the word was derived) so the word supervisory refers to the **SCADA** only changing setpoints and going from auto to manual.

RTU
Remote Terminal Unit
A local control device and has a CPU or **PLC**. The term is typical used when the control device is located far away from the **SCADA** computer. Usually connected via radio, phone lines, microwave etc.

PLC
Programmable logic controller
Used to control plant process both analog and digital. Also used for the field **I/O** connection to get data to the SCADA system.

DCS
Distributed Control System
Larger and more powerful (lot more cost) control system than **PLC**'s. Usually only used when many analog loops are used for control. Very rare in Water Plants but they do exist there.

Tag Name
Generic term for the I/O point description name in a SCADA system. Most brands use this term.

I/O
Input and Outputs

Protocols
The software language that is used for communications between equipment. Some are open and some are proprietary Ethernet is a protocol. An example is 1. Allen Bradley PLC's use DF1, DH and DH+, Modicon PLC's use Modbus and MB+, TI uses Tiway etc.

APPENDIX C
REQUIREMENTS ANALYSIS FOR THE UTILITIES
CONTENTS

TABLES

FIGURES

INTRODUCTION

The drinking water industry has long been rich in data but poor in information derived from these data. Routinely, utilities perform numerous measurements related to the management of water resources and the quality of source water, finished water and water delivered to customers. Utilities measure a wide range of data through a variety of monitoring strategies. Field technicians and operators make in-situ measurements, while automated measurements collected via Supervisory Control and Data Acquisition (SCADA) Systems are made using inline sensors. Grab samples are taken from diverse locations and are then analyzed in utility and commercial laboratories. Hydrologic, weather, demographic and emergency data, critical to effective management of drinking water utilities, are collected and disseminated by municipal, state and federal agencies. Much of the data described above are used daily to make operational decisions regarding water resource management, water treatment, and distribution strategies. Failure of accurate and timely reporting of some of these data can result in compliance violations, which can be costly and embarrassing for utilities. If managed, aggregated and stored effectively, these data could help determine management strategies for repeated situations.

A broad understanding of the operations and management of utilities is needed in order to catalog the information to support decision processes for optimal operation of a drinking water utility. To develop this broad understanding, the project team is conducting requirements analyses aimed at determining data sources, data processing, reporting, and information users within a number of utilities. These utilities were chosen to represent a broad range of geographic distribution, source water types, treatment types and levels of sophistication in the use of information technologies. The participating utilities are listed in Table C.1.

The effective use of data and information requires an understanding of data quality and aggregation methods that have been utilized. Figure C.1 illustrates the conversion of data into information and knowledge and its relationship to prospective users from technicians and operators through to the public. As data are processed, analyzed and migrated up the left triangle, decisions are made regarding aggregation strategies, data exclusion due to quality assurance issues and statistical approaches. Therefore, information systems content require increasingly more documentation in order for content to be used effectively. As a general trend the users become less indoctrinated in the detailed uses of databases and computer systems as the categories go up the right hand triangle requiring more aggregations, simpler interfaces and concise summaries of the documentation.

Table C.1
Utilities involved with the development of a consolidated reference of water quality management and decision-making

Utility	Location	Population served	Source Type	Treatment Type
Massachusetts Water Resources Authority	Massachusetts	2,500,000	Surface Water	Chemical
City of Golden	Colorado	16,200	Surface Water	Conventional
City of Tucson	Arizona	600,000	Ground Water	Chemical
East Bay Municipal Utility District	California	1,200,000	Surface Water	Variety
City of Newport News	Virginia	360,000	Surface Water	Conventional

In order to develop a comprehensive understanding of water quality management and decision making processes in drinking water utilities, the project team conducts focus group workshops involving all divisions within a utility that either generate or use water quality data in management and decision making. The workshops focus on the types of information generated, the processing of the data and samples, the aggregation and interpretation of the data that is necessary, and the decisions based on the data. The different levels of data and users shown in Figure C.1 are evaluated, and the content of information required for each level of decision making captured. This evaluation is performed from a variety of perspectives including current practices, planned upgrades and visions of how the utilities would like to manage and apply data. The evaluation examines decisions on a wide range of scales, including an evaluation of the information, which is collected, current collection methods, information management and use in decision making, as well as its use in improving decision making.

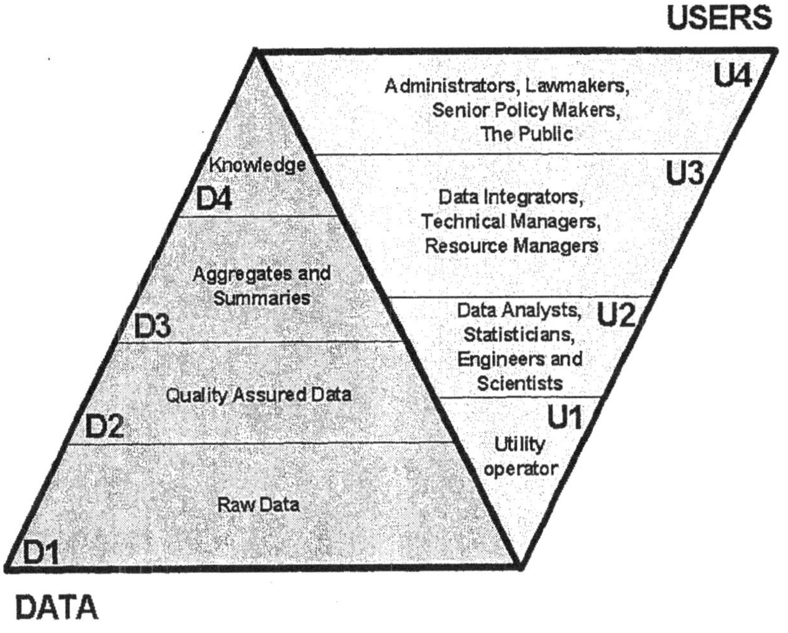

Figure C.1

Pyramid indicating the users of information as it becomes more processed

Some critical data related to water quality, operations and management of a utility are gathered and maintained by unexpected departments in a utility. For example, much of the critical data on water usage is often found in the billing system since meter readings for individual customers are entered into this system. Similarly, the uses of water quality data and their aggregates are extensive throughout a utility. Therefore, in order to ensure that the project team understands all aspects of the uses of water quality data, an assortment of divisions, departments or functional areas are covered in the focus group meetings.

The project team focuses on the following groups or functional areas within each utility:

- Water Quality
- Laboratory
- Water Treatment
- Hydrology
- Operations
- Maintenance

- Planning and Engineering
- Information Services
- Administration
- Senior Management

Each utility can be organized differently and these functional areas can cross organizational boundaries. The project team works with each utility to define the focus groups that need to be convened. The team focuses on generators of water quality data with special focus on SCADA systems, field monitoring and laboratory analyses. Other key sources of water quality data are identified and documented.

There are several key steps involved in the on-site analysis of a utility. Once the key stakeholders have been identified to represent the functional areas described previously the following steps are taken:

a. Questionnaires are distributed to the stakeholders to help the project team develop a good understanding of the information needs and organization of the functional areas. This also helps prepare the participants for the focus group meetings.
b. Focus group meetings are held with each functional area individually to develop an in-depth understanding of the area's requirements and current practices. Current practices for data management and processing are documented along with planned improvements. Also, barriers to effective data access are explored and documented.
c. The project team summarizes the findings from each utility and returns it to the participants for review and comment.

The following sections contain the five utility evaluations. They are presented according to size starting with the smallest utility, City of Golden, followed by Newport News, Tucson Water Department, East Bay Municipal Utility District, and Massachusetts Water Resources Authority.

CITY OF GOLDEN

UTILITY MANAGEMENT

The City of Golden, Colorado is located 10 miles west of Denver at the east end of Clear Creek Canyon. About 16,000 people live in Golden. Within Public Works, there are seven departments that employ approximately 35 people. The Water Treatment Plant and Environmental Services are the two departments responsible for the water quality and treatment plant operations data. The Water Treatment Plant has a plant manager, six full-time operators and during the summer one part-time operator. The Environmental Services Department has five employees including two chemists and three people responsible for special projects.

The Distribution System Operations department is responsible for the maintenance of the distribution system and assists the Water Treatment Plant Superintendent. Together with the manager of the department, nine field personnel are employed in the Distribution System Operations department. The Engineering Department is responsible for all City construction projects, serving the Water Department by designing and overseeing the construction of distribution system piping. The department consists of a City Engineer, two project engineers, and a construction manager. The GIS function for the City is currently housed within the Public Works Department and consists of a single employee. The Information Technology Department is a city-wide department housed outside of Public Works, but provides services to the Public Works division and currently employees three people when fully staffed and will likely expand to four full time employees in 2001.

WATER TREATMENT PLANT OPERATIONS

The City of Golden (City) operates a 15 MGD conventional treatment plant. Clear Creek (source) is predominately mountain snowmelt. Flow from the creek is diverted through a series of two ponds. From the ponds, it is pumped to the headworks during which potassium permanganate, ferric sulfate and lime are added. Flow is typically split at the rapid mix with approximately half the flow going to the south basin and half to the north; however, each side can be isolated for cleaning. Following rapid mix, flocculation aid polymer is added as the water flows through three-stage flocculation. Plate-aided sedimentation followed by rapid filtration (mixed media) completes the process train. Chlorine is used for disinfection.

Cold water temperatures during the winter and high metals concentrations from mine drainage challenge the plant. In addition, organics flushed from forest floors into the creek during spring runoff can impart high TOC and taste and odor events. Manganese, the most challenging metal, is controlled by preoxidation with potassium permanganate as well as adsorption onto the filter media. Some success in controlling taste and odors has been observed while conducting testing for enhanced coagulation. Organics removal is achieved through coagulation. This process is still being fine-tuned in preparation for compliance with the Stage I D/DBP rule.

PRELIMINARY DATA COLLECTION AND INTERVIEWS

Twelve questionnaires were completed and returned to the project team prior to their site visit. The interview and reporting process for the City of Golden was scheduled over three days, July 19 through July 21. The interviews were conducted all day on the first day. The project team had identified eight groups and each was interviewed for approximately one hour. These groups included:

- Plant Operations
- Environmental Services
- GIS/IT
- Maintenance
- Metering
- Plant Management
- Water Quality
- Distribution
- Engineering

On the second day, the project team met and summarized their findings from the previous day's interviews. A presentation of the data collected and findings were presented back to the City staff on the third day. This was done in an effort to validate the information collected, facilitate further discussion and present preliminary results of the information collected. Table C.2 lists the participants in the focus group meetings as well as the final day presentation.

Table C.2

Stakeholders

Name	Department	Questionnaire	Interview	Presentation
John McEncroe	Plant Manager	X	X	X
Anthony Sotelo	Water Treatment	X	X	X
Shari Matousek	Water Treatment		X	X
Greg Demhlow	Water Treatment	X		
Becky Hull	Env. Services	X	X	X
Vicki Coppage	Env. Services	X	X	X
Fran MacLean Phillips	Env. Services	X	X	X
Tracey Pond	Env. Services	X	X	X
Greg Moore	Env. Services	X	X	X
Greg Mortimer	IT	X	X	
Quint Pertzsch	GIS	X	X	X
Byron McElhinney	Maintenance	X	X	X
Terry Tilkes	Superintendent		X	X
Tony Warner	Crew Leader		X	
John Green	Metering		X	X
Scot Lewis	City Engineer	X	X	X
Les Major	Inspector		X	X
Dan Hartman	Public Works Director			X

DATA STREAMS

Water quality and operations data is collected by the water treatment plant staff and the environmental services staff. A summary of the major data groups, their uses and storage is summarized in Table C.3. Data is typically collected via online or grab methods, and it is stored either electronically or in hardcopy form. It is common for data to be gathered by the SCADA system and then transferred to hardcopy or manually collected and entered into an electronic format. For example, data collected for turbidity compliance reports is taken from the SCADA system and written on a data sheet that is sent to the state. Data collected to confirm manganese removal through the filters is first analyzed in the lab and then entered into spreadsheet and database programs.

Table C.3

Information and Data Generation

Department	Type of Information	Collection	Storage
Water Treatment Plant	Process Water Quality (i.e., metals, particles)	Online, Grab	Electronic, Hardcopy
	Operation Parameters (i.e., flow, chemical feed rates, tank levels)	Online, Grab	Electronic, Hardcopy
	Compliance Parameters	Online, Grab	Electronic, Hardcopy
	Emails, Notes, Log	PC, Grab	Electronic, Hardcopy
	Reports	NA	Electronic, Hardcopy
	SOPs	NA	Electronic, Hardcopy
Environmental Services	Process Water Quality	Grab	Electronic
	Compliance Parameters	Grab	Electronic, Hardcopy
	Emails, Notes, Complaints Log	PC, Phone	Electronic, Hardcopy
	Reports	NA	Electronic, Hardcopy

Numerous sources and users of information and data were identified. In addition to the plant and lab data, the City's billing system, maintenance management program, GIS data and e-mail were identified as major data sources. Through the interview process a number of internal and external users of the generated data were identified (Table C.4). Tables C.5 and C.6 summarize the electronic data sources and the related internal and external data users.

Table C.4
Information and Data Users

Users within Public Works	External Users
❑ Operation	❑ Regulatory Agencies
❑ Maintenance	❑ Other City Departments
❑ Distribution	❑ General Public
❑ Environmental Services	❑ City Council
❑ Engineering	❑ Consultants
❑ GIS	❑ County Agencies
❑ Finance	❑ Organization/Business
❑ Management	❑ Special Projects

Table C.5
Data Sources and Internal Users

Major Sources of Electronic Data	Major Internal Users (Current)							
	Operation	Maintenance	Distribution	Environmental Services	Engineering	GIS	Management	Finance
SCADA	X	X	X				X	
Laboratory	X		X	X	X			X
Billing System			X	X				X
Maintenance Management			X	X	X	X		
GIS			X	X	X	X		
Email	X	X	X	X	X	X	X	X

Table C.6
Data Sources and External Users

Major Sources of Electronic Data	Major External Users (Current)							
	Regulatory Agencies	Other City Departments	General Public	City Council	Consultants	County Agencies	Organizations /Businesses	Special Projects
SCADA	X	X	X	X	X		X	X
Laboratory	X	X	X	X	X	X	X	
Billing System		X	X	X	X		X	
Maintenance Management		X	X	X	X	X		
GIS			X	X	X	X	X	
Email	X	X	X	X	X	X	X	X

INFORMATION TECHNOLOGY

The City of Golden Water Department is part of the city infrastructure. Currently, the city has a single IT person supporting the entire infrastructure. They are in the process of hiring two additional people. One person had recently left reducing the staff to one person. This is inadequate given the requirements that are specified below.

The City of Golden, Colorado has an IP based network consisting of 17 servers, three of which are Windows NT platforms supporting the Water Department. Other servers in the city also include four UNIX boxes and an AS/400 that is used by the public safety group. The Water Department uses Microsoft office tools throughout the organization. Many desktop machines are located throughout the organization both within the treatment as well as distribution systems, laboratories and other facilities within the Drinking Water Facilities. The City of Golden Water Department has a newly initiated GIS effort underway which employs a suite of ESRI tools (ArcInfo, ArcView and others). Eventually, they plan to move into MapObjects and Internet Map Services (IMS). Environmental Services, which performs water quality analysis, currently stores their data using a software package called TRIBAL. TRIBAL is a laboratory information management system (LIMS) that is written on top of Microsoft Access and can connect with a database in SQL Server. The database includes the required hooks for GIS interfacing: specifically latitude, longitude, altitude and station ID. The current SCADA system is a DOS-based package called GENESIS. Other information systems include a maintenance management system, Cartograph, used by the Distribution group but not the Plant Maintenance engineer.

SCADA SYSTEM

The City of Golden presently has a Supervisory Control and Data Acquisition (SCADA) system in place for control and data acquisition. Twelve Bristol Babcock processors perform the control logic for plant control and process the field inputs and outputs. Six of the processors are located at the remote locations and communicate to the plant via 1200-baud rate modems. The host computer that collects the data for operator interface is a DOS-based system manufactured by Iconics (Genesis). The system has 1200 data inputs and outputs that are both digital and analog data. The system is capable of trending 40 analog inputs at 12-min intervals. The trend data is archived monthly on a floppy disk. The system in place is obsolete and Golden has allocated funding to upgrade the system. The present system is operational; however, the alarm printer is not functional at this time due to the new printers only have Windows based drivers. The system currently "crashes" several times a week and has to be re-booted. The system presently has very limited report capability.

OBSERVATIONS

Each of the core information systems currently employed at the Golden Water Utility is focused in the hands of a single operating division. For example, the distribution system personnel currently use the Cartograph system with little use by other employees and departments. The GIS system is currently used predominantly by the GIS person (Quint Pertzsch). The LIMS is predominantly used by the Environmental Services group. In all cases there are few users from outside a given division that use the division's information systems. The majority of data exchange is done through personal communications in which a person will ask one of the people within the division to produce a specific report for them. This approach has been adequate for Golden because there are very few people employed and they all get along very well. The majority of the data exchange that occurs is done on paper, but often that data has been extracted from an electronic information system. There is very little information available on centralized data base systems that are available via the local area network. There are clearly extensive opportunities for expanded use of data within the Golden Water Utility and the City of Golden in general. There is currently a major trend underway within Golden to computerize information processing and availability. The development of a Geographic Information System has just been initiated and already is making major strides in a most productive direction. Very little resistance was exhibited within the utility to the use of the GIS or to the standardization of naming conventions both for sampling points and for specifying latitude and longitude. Most of the employees appear excited about the potential for GIS within Golden.

The utility is currently involved in a number of significant information startups, specifically, GIS, Cartograph and TRIBAL. These startups are potentially tapping the key information management resources within Golden. The same people who are attempting to initiate the startups also maintain daily responsibility for operations and processing of standard data streams. It appears that the startups are significantly hampered by the lack of additional temporary resources. Whereas these limitations are temporary for the laboratory, Cartograph and a few other efforts, they are longer term for GIS and general IT support.

The primary use of the SCADA system is for Operations to monitor and control plant functions. Other uses include the maintenance staff, which utilizes the system to aid in trouble-shooting plant problems and the distribution staff, which utilizes the system to gather distribution data such as tank levels to assist in scheduling maintenance. Other departmental groups do not have access to online or historical SCADA data. The data from the SCADA historian is not transferred to a relational database for use in reports and is not shared with any other computer system. The City of Golden utilizes contract labor for support of the SCADA system.

BARRIERS

The following barriers prevent the effective transfer of information data within the Golden City Water Utility:

- The utility is paper centric. The initial information capture within the utility is predominantly in a paper form with the exception of the SCADA system. This paper centric approach stifles conversion to electronic systems since it requires additional tedious work. The paper centric approach makes it difficult to track, locate and share information that is required. This paper centric preference is not due to discomfort with electronic transfer of data but does have some roots in a lack of confidence in the authoritative records being electronic versus paper. This is based on history in which the authoritative records on any transactions in the past have been on paper.
- There is a lack of a central repository of information. This lack of a central repository prevents access to on-line data. Examples of important useful information that has limited accessibility on-line include billing and water quality (outside of Environmental Services).
- Currently, the utility workers are not aware of much of the external data that is available to them. Some people are aware of external sources of data like USGS gauging stations and STORET; however, others within the utility are not familiar with accessing this information. To some degree, locating and accessing this information would be simplified through training in the use of these systems.
- The utility cannot submit its results and compliance reports electronically because the state's regulatory agencies are currently not prepared to accept electronic records.
- Critical utility information that is not currently paper-centric is people-centric with a large array of critical information for the utility maintained in the heads of a few critical people. If for any reason these people were to leave or were no longer

available for work, much of the critical information available within the utility would disappear.

- There is currently no single person within the utility who is responsible for the integration and knowledge of available information resources.
- The common vision of moving towards computerized information systems and making this information available over a network has not been adequately communicated to the utility community.
- The utility is resource limited, specifically in personnel, training and funding.
- Some limitations were also identified relative to communications. These limitations were expressed as being a result of lack of time rather than lack of will to communicate.
- The utility currently does not have adequate field systems for accessing and entering information in the field where samples and data are collected.
- There is currently a lack of historical records about the utility infrastructure and GIS base maps for the city's infrastructure (i.e., pipes, valves, pumps, boosters, etc.) making it difficult to perform change analyses.
- There is a significant need for a scheduling module that can be used to determine who is available for training, communications, etc. The schedule module will also be useful to coordinate the work of different departments. Examples were given of Engineering and Distribution knowing when work will be done out in the streets so that it can be coordinated. This would eliminate the possibility of one department ripping up the streets to put in new pipes and the other department coming out two weeks later to do other work again requiring construction of the street.

REQUIREMENTS

This section summarizes the major requirements that were articulated by the focus group participants and later in the joint requirements planning meeting.

Few requirements were identified for the exchange of information between Golden Water and the Public Works Director. Significant information exchange is needed between the Water Utility and the following other city departments:

- Engineering
- Fire
- Police
- Parks
- Finance

The utility needs to regularly exchange information with regulatory agencies, the public, city council, consultants, county agencies and other organizations.

The utility has a need for the capture, aggregation and presentation of a large amount of information from its current and planned upgrade SCADA system.

The utility needs to track samples analyzed both within house and sent out to contract laboratories and to correctly associate the results of those analyses with the original samples. This information is required in a computerized form that can be interpreted, aggregated and built into compliance reports.

The facility has major needs for the development of compliance reports that can be distributed to agencies at the state, county and federal level.

The utility interacts extensively with the Coors's Brewery and needs to both receive as well as send information to the Coors's Brewery regarding treatment of both freshwater and wastewater.

The utility currently relies heavily on paper records of a wide variety of formats including but not limited to: data sheets, operator observations, blue prints, as-built drawings, permits, reports, contract lab results, letters, work assignments, engineer notes, inspections, contracts, bond documents, etc. These and other paper records are currently not tracked in any formal way.

A widespread need for geospatial data was expressed. These needs include standard location names, distribution system infrastructure, contours, elevations, customer locations, customer classifications, water resources, valves, etc. Basic geospatial data coverages are needed to act as a base for developing a comprehensive spatial database for Golden Water.

Improved scheduling and coordination between departments and groups is needed to reduce conflicting maintenance.

Scheduled maintenance is limited in the plant, but not in the distribution system. In many cases, standard operating procedures are not documented resulting in inconsistencies in approach. A scheduled maintenance system that includes scheduling and Standard Operating Procedures needs to be implemented.

Summary information regarding the operations, performance and problems of the previous day would be helpful especially if the record was on line and cross-referenced to pertinent data and information.

RECOMMENDATIONS

The following recommendations result from integration of all the information gathered during the focus group meetings.

- **Golden Water should consider the installation of a records management system.** Records management would be instrumental in tracking paper documents including blue prints, as-builts, and other critical paper, which is currently a primary path for information communications.
- **Move to a centralized data system for access.** This system would take output at some aggregated levels from the other key systems within Golden Water and make this information available over the LAN to all employees within the Water Department and possibly other city departments. Some subset of this centralized database would also be further aggregated and made available to the public over the public website.
- **Schedule, coordination and management system.** Golden Water should institutionalize the use of a scheduling system. The scheduling system within Microsoft Outlook would probably be an adequate start for the department.

- **Contract supports for startups.** Golden Water should plan on hiring contractors to help the initiation of the **TRIBAL** Laboratory Information Management System, **Cartograph** Maintenance Management System, and the **Geographic Information Systems** efforts. The plans for the contractual support should include in detail, transition to standard operation by Golden Water personnel.
- **Additional full-time positions for GIS and Information Technology Support.** The City of Golden and the Golden Water Department should plan on additional full-time equivalent positions supporting Geographic Information Systems and information technology. Depending on the long-term plans for the Geographic Information Systems, it is likely that one and one-half to two additional FTEs will be required. In the IT area, the City needs to fill its two outstanding positions and might consider an additional position for information organization and distribution (the Information Librarian). In previous studies, TPMC has noted a need for an information bureau. The City of Golden is too small for an information bureau but does need someone to focus the information resources within the City.
- **Expand sources for base GIS coverages.** The City of Golden should expand its excavation for GIS coverages. By this, we mean that many GIS coverages may already exist that could be accessible to the city either free or through purchase. Sources currently being explored include the county; however, other sources may be available including other utilities such as gas and National Clearing Houses.
- **Formulize standards.** The Golden Water Department has begun institutionalizing standards. These should be clearly documented and formalized for everyone to access and reference.
- **Document standard operating procedures.** Similar to the bullet above, Golden Water should document all standard operating procedures. This would focus on approaches to plant operations, maintenance and distribution system operation maintenance. This would reduce the dependence on individual people. This effort might also require contractual support. These SOPs should be available on-line either as part of the maintenance management system or as a simple locked directory with word processor or PDF files.

- **Development a preventative maintenance system.** Preventative maintenance is currently done in an ad hoc form or not at all. It is likely that the Cartograph system contains a preventative maintenance module. This should be employed by the distribution system and by plant engineers.
- **Develop field information systems.** Golden Water should begin exploring the use of Palm Pilots and hand-held computers for use in the field. These efforts should focus on making information from central databases available to field personnel (example would be the availability of metering information on a Palm Pilot for the meter support team). This should also be explored for data entry including input to the Cartograph system as well as GPS readings. This effort should begin with a small pilot effort.
- Knowledge management is a long-range option but not needed at this point.
- **Install SCADA connections to the meter at the insolent to the Coors's wastewater treatment plant.** This is an important source of critical information to the laboratory and much consternation regarding compliance reporting could be alleviated by installing a connection to this meter and making the information available to the central information management system.

SCADA SPECIFIC RECOMMENDATIONS

The new SCADA system should be specified to transfer the historian data to a relational database such as Access. This will facilitate the data to be used with other computer systems even if it is not an automatic transfer. The new SCADA software should be capable of transferring data to other computer systems using ODBC, SQL and DDE transfers. The SCADA software should be able to directly import trend page numerical data into Microsoft Excel using a "click and drag" method. This allows for data to be captured in an open system format without reports being generated. The SCADA software should use object based technology and provide for Active X controls acting as an Active X container. At least one of the host computers should be supplied with a "development" license package that allows for configuration and graphic changes. This will ensure that plant personnel and contractors can make changes to the system without buying additional software. A "view only" terminal should be installed in the laboratory room. This would allow sample data to be entered into the SCADA system for online trend comparisons and allow lab personnel to see on-line instrument data. The new SCADA system should be supplied with a remote paging software. This will allow for critical alarms that are not acknowledged in a timely fashion, to send an alarm page to a paging system or telephone system. Information from SCADA should be exported to a common database with other systems such as LIMS, Billing, GIS, and Maintenance Management.

The present plan to upgrade to a new version of Bristol Babcock will limit the amount of contract labor available to support the control system. The new SCADA system should be based on "open" PLC technology to allow for future plant upgrade connectivity and provide for more contract support. The Bristol Babcock system is used primarily in the water industry. The PLCs are used in every industry including water and therefore provide a more open system to support. The future process equipment upgrades will come with PLC equipment and can easily be added to the SCADA system without major upgrades or cost. The PLC programming software should be Windows NT compatible.

NEWPORT NEWS WATERWORKS

UTILITY MANAGEMENT

The City of Newport News, Virginia is located at the junction of the James River with the Chesapeake Bay in eastern Virginia (Figure C.2). The City's Waterworks serves nearly 400,000 people residing in Poquoson, Hampton, Newport News, and portions of York and James City counties. The Waterworks employs approximately 370 people within its seven functional divisions under the Office of the Director. The functional divisions are:

a Planning and Programs Division
b Engineering Division
c Finance and Administration Division
d Facilities and Watershed Maintenance Division
e Water Production Division
f Distribution System Operations Division
g Material and Equipment Maintenance Division

The key divisions involved in this assessment of information management were:

- ***Office of the Director***. Responsible for all aspects of the Waterworks operation. The Office of the Director includes the Director, Assistant Director, and Administrative Manager for the Waterworks as well as administrative support. The Waterworks' GIS initiative operates under the Director's Office as well.
- ***Finance and Administration***. Responsible for billing, collections and business administration.
- ***Engineering***. Responsible for the engineering design and construction of all transmission and delivery systems.

- ***Facilities and Watershed Maintenance.*** Responsible for the maintenance of all water production facilities and activities within the watershed area controlled by the Waterworks.

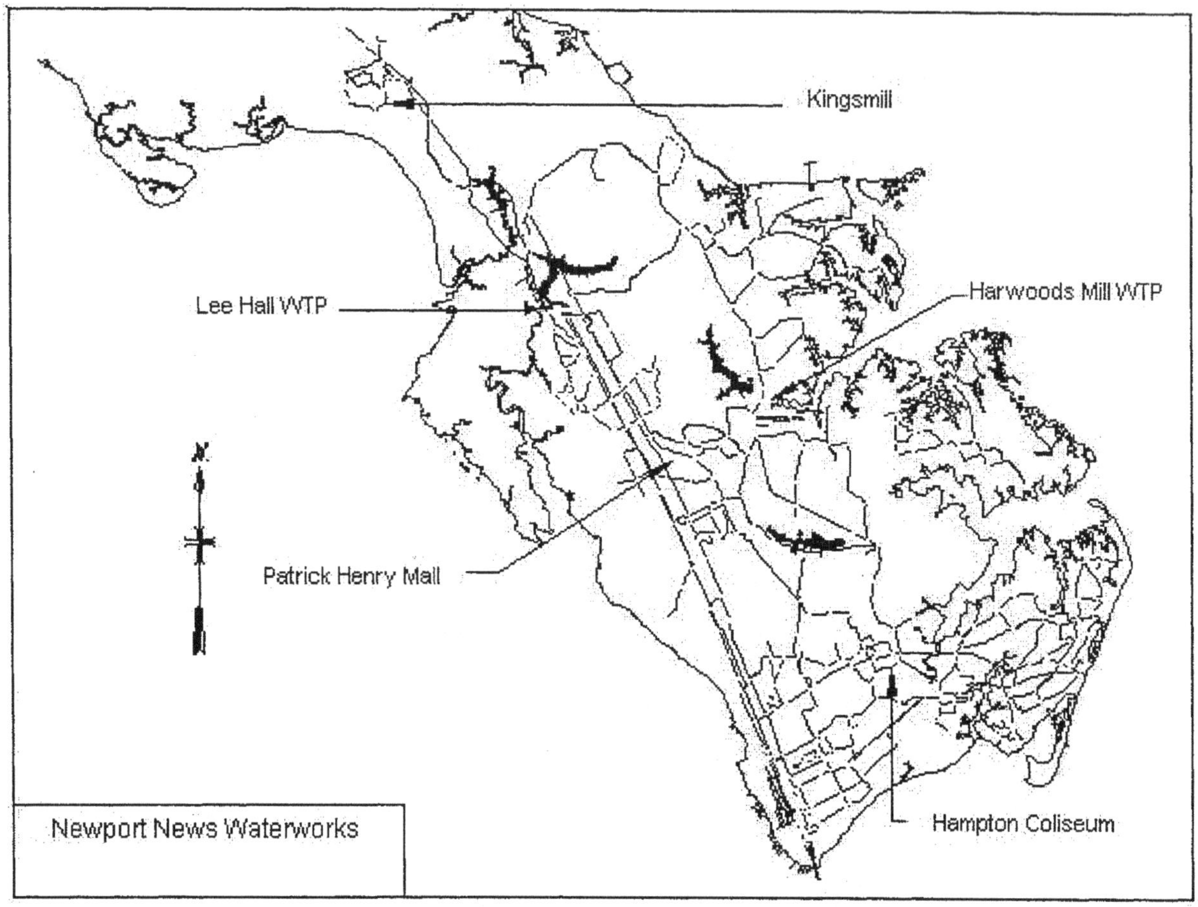

Figure C.2
City of Newport News

- ***Water Production.*** Responsible for all aspects of water production. Within this division are such departments as water treatment plant operations, laboratory services, and process control specialists. Also responsible for delivering the water under normal operational conditions.
- ***Distribution System Operations.*** Responsible for the installation, inspection, repair, and maintenance of the distribution, transmission, and related systems

WATER TREATMENT PLANT OPERATIONS

The City of Newport News (the City) operates three water treatment plants, two conventional surface water plants and one reverse osmosis groundwater (brackish) plant. In addition to three treatment plants, Newport News currently operates 10 pumping stations, 12 tanks, and is responsible for 8,500 acres of watershed. The Lee Hall plant was originally built in 1907 with significant upgrades in 1967 and 1973. It is a completely manually operated conventional plant with a rated design capacity of 54 millions of gallons per day (mgd). Some water quality and operational measures are collected using inline sensors with the measures integrated with the City's SCADA system. A new replacement plant is currently under construction and will be a fully automated conventional plant with ozonation. The Harwood's Mill plant was completed in 1989 and was designed as a fully automated conventional facility with a rated design capacity of 31 mgd. Currently, some of the operations are still operated manually. The groundwater facility was brought on-line in 1998 and treats brackish groundwater through a fully automated reverse osmosis process with a rated capacity of 6 mgd. In addition to the full-scale facilities, the City has constructed and operates a pilot treatment plant that has inline data collection integrated into the City's SCADA system.

In addition to the water treatment facilities, the City also operates a centralized residual handling and disposal operation and a watershed management program. The residuals produced by the Lee Hall plant are pumped to a centralized centrifugation process located at the Harvey Mills plant. The resultant solids are then trucked to a land application operation owned and managed by the City within the Waterworks' watershed. Groundwater monitoring wells are routinely sampled to assure the quality of the underlying groundwater system. The City also monitors the composition of the applied residuals and soils, tree growth, and vegetation species located in the land application area and other areas of the watershed controlled by the City.

The City's water demand averages 45 mgd and remains fairly constant on a year-round basis due to the large amount of precipitation this region receives. Each of the plants is base-loaded in terms of production, and peaking operations are unnecessary. The water quality received by the two surface water plants, though, presents certain treatment challenges. While the plants currently provide conventional treatment (coagulation, flocculation, and filtration with chlorine/chloramines disinfection), the new Lee Hall plant will have ozone treatment and the

Harvey Mills plant is being retrofit with ozone treatment facilities. The high organic content of the source water to these facilities has led to the production of high disinfection by-product levels in the distribution system. Based on early pilot studies, the use of ozone to meet the disinfection and oxidation demands of their supply results in an overall reduction in DBP levels. The final disinfectant for the distribution system will remain as chloramines.

The water treatment plants currently monitor for turbidity, particle counts, pH, and chlorine residual through inline sensors. Inline sensors are also used for such physical parameters as flow rate, reservoir levels, precipitation, chemical feed rates, and valve settings. The inline monitoring data are communicated to the SCADA system, which stores them compactly based on a "change of state" for 14 months, after which the measures are downloaded for permanent archives on tape. The inline monitoring data are of substantial use to the City's Water Production Division in a non-operation mode for research and special studies off-line mode (exported to spreadsheets or other databases). In fact, the requirements for the frequency of data collection and measurement resolution by inline sensors are driven by these studies more so than the system control requirements.

PRELIMINARY DATA COLLECTION AND INTERVIEWS

Seven questionnaires were completed and returned to the project team prior to their site visit. The interview and reporting process for the City of Newport News was scheduled over three days, September 18th through the 20th. Interviews with Waterworks staff were conducted all day on the first day and through the morning of the second day. Each questionnaire was completed for the representative group within the Waterworks with which interviews were conducted, namely:

- Water Production - Operations
- Water Production - Laboratory
- Water Production - Process Control Specialists
- Facility and Watershed Maintenance
- Water Information Systems and GIS

- Engineering
- Distribution System Operations
- Office of the Director

On the remainder of the second day, the project team met and summarized their findings from the various interviews and preliminary questionnaires. A presentation of the information collected and findings was made to the City staff on the third day. The goal of this reporting out was to validate the information collected, facilitate further discussion and present preliminary results of the information collected. Table C.7 lists the participants in the focus group meetings as well as the final day presentation.

Table C.7

Stakeholders

Name	Department	Questionnaire	Interview	Presentation
Tom Kahler	Facility Maint.	X	X	X
Chris Harbin	Water Production	X	X	X
Robert Barclay	Water Production		X	X
Danny Pratt	Water Production		X	X
Mike Hotaling	Water Production		X	X
Karen Levy	WIS	X	X	X
Donald Joyner	WIS	X	X	X
Charlene Sevier	GIS	X	X	X
Marietta Washington	GIS	X	X	X
Chris Basford	GIS	X	X	X
Jim LaVasseur	Operations		X	X
Bryan Bradish	Engineering	X	X	X
Billy Forbes	Engineering		X	
Daniel Kuhns	Operations	X	X	X
Eileen Leininger	Director's Office		X	X
Sherry Williams	WQ lab	X	X	X
Jeff Pletcher	Water Production	X	X	X
Nancy McNelly	Water Production		X	
Chris Gillis	Water Production		X	X
Bruce Utne	Water Production		X	X
Darlene Mormon	Customer Service			X
Billy Forbes	Engineering			X
Kris Keyes	Director's Office			X
Tom Murphy	Finance & Admin.			X
LeeAnn Sink	Public Info			X
Donna Hautz	Facility Maint.			X
Scott Dewhirst	Water Production			X
Brian Ramaley	Director's Office	X		X

INFORMATION TECHNOLOGY

Maintenance of the plants is managed by a computerized maintenance management system (CMMS) called ORION developed by Lifecycle Engineering in Charleston, SC. This is a system, which was built specifically for Newport News in 1993 based on an existing system developed internally at the Waterworks in 1987. The CMMS interacts with the City's public utility billing system (PUBS) to capture some of the information regarding required maintenance. The actual work orders that are acted upon in this group are generated out of the PUBS system on the City of Newport News's mainframe. Materials management is handled by another mainframe system, ICMS. All of the man-hours and materials costs are maintained on the mainframe whereas ORION maintains the maintenance history and the scheduled maintenance. The actual schedule for maintenance is output from ORION. Any problems regarding the completion of work orders are documented on the mainframe system. Direct access to ORION is limited to plant maintenance personnel, but snapshots of data are distributed via the Waterworks' Intranet discussed next.

Newport News Water has developed an intranet system, WaterNet, which contains summary data and reports. For example, snapshots of information from both the Orion and SCADA systems are published on WaterNet. WaterNet is populated with outputs from other systems as well. This Intranet is centralized and available to any Newport News Waterworks employee. The information contained in this repository is predominantly generated by analyses and integrations performed by the Waterworks Information Systems (WIS) group. WaterNet is currently not directly linked to any other database and there is no active search of other databases from the WaterNet System (the system is basically static). WIS plans to move quickly to dynamic access of data from WaterNet. This will be facilitated using standard commercially available software. WIS expects to allow subsetting of the available data and ad hoc queries and report generation.

The WIS group is a water department funded group but it interacts extensively with the City of Newport News IT department. The Waterworks transfers funds to the City of Newport News each fiscal year to support three programmers on the City IT staff. These programmers are predominately COBOL programmers who are implementing changes and maintenance on the legacy systems that run on VSAM files on a mainframe. In general the WIS group acts as the clearinghouse for information within the Waterworks department.

The Waterworks has access to a website that is maintained by the Data Processing Department for the city. The Data Processing Department maintains the firewall and also acts as the Internet service provider for all city personnel and departments. The WIS group is responsible for the entire Waterworks backbone and all of its security.

Major software systems used within Waterworks include: Microsoft Office 97 Suite, they use Outlook 98, Internet Explorer 4.0, IBM personal communications/3270 (link to E server), Adobe Page Maker 6.5, Norton Antivirus - Corporate edition 7. The Engineering Group also uses AutoCAD 2000, CAD Overlay 2000, and Cyco Auto Manager workflow. Geographic Information Systems are done predominately using ESRI tools. All data coverages, shape files and electronic maps are being accessed using ArcView and ArcExplorer. They are using Pathfinder to convert GPS unit output to GIS importable files. The GIS group plans to move towards the use of both ArcIMS and SDE but is not currently using these tools.

WIS has been working closely with the field maintenance personnel to initiate field-computing capabilities. Currently there are five laptop computers that are in the field using Cellular Digital Packet Data (CDPD) technology to communicate wirelessly. This reduces voice radio traffic and also reduces the need for personnel in the field to go back to headquarters to gather information. The success of this has generated the order of six more laptops.

The WIS Group is working closely with the city mainframe to develop front ends for existing information systems using 3270 terminal emulation and relational database technologies. The Waterworks currently has 193 gigabytes of storage capacity on 26 separate hard drives, but not all of this capacity is available for storage as some of the servers and hard drives have very specific roles, including acting as fax servers, backup servers, and other dedicated activities.

SCADA SYSTEM

The City of Newport News currently has a Supervisory Control and Data Acquisition (SCADA) system in place for control and data acquisition. The system is about three years old and is utilized extensively by operations personnel. The Human Machine Interface (HMI) is manufactured by HSQ (Miser software) and the operating system is VMS running on DEC Alpha Servers. The HMI has 4500 Tags (inputs and outputs) and the Miser software collects historical data for 800 points. The Miser software has drivers to allow an export to an Oracle database, but is not yet used by the Waterworks. The controllers in the water plant for collecting field inputs/outputs and for process control are Allen Bradley PLC 5 and SLC 500 communicating Ethernet to the to HMI. The 15 remote sites have HSQ controllers communicating Ethernet via frame relay to the HMI. There is one GE PLC for generator control communicating Ethernet to the HMI. The Hardwood Mills and Lee Hall plants are networked together via a Fiber WAN with one of the servers operating as the primary server and the other as a "hot standby" server. External users have access to the "real time" and historical data via DEC excursion (X-widows).

The SCADA system is currently set up to record information about any of its tag points (monitored locations) based on a change of state. A change of state is defined as a significant change in the measurement. The magnitude of change required to trigger an entry in the database is set on the SCADA system by the plant managers. The plant managers indicated that the collection of data is driven to a large degree by research, not by operations. A great deal of information is maintained regarding operations as well as turbidity, particle counts, pH, and residuals chlorine concentrations, but these data are seldom revisited with the exception of the data required for research. All the data generated by the SCADA are currently stored and maintained on the separate SCADA servers. Currently, the SCADA system houses up to 14 months of data on-line. Additional archived data is available offline on tape and can be reloaded as necessary. The archived data are seldom used for operational purposes but have been reloaded for some research projects

DATA STREAMS

Water quality and operations data are collected by the water treatment plant operations and laboratory staff. A summary of the major data groups, their uses and storage is presented in Table C.8. Data are typically collected via on-line or based on analyses of grab samples, and it is stored either electronically or in hardcopy form. When the City upgraded its SCADA system (1999), it selected a SCADA system that would enable the entry of grab sampling results in order to consolidate on-line and grab monitoring data into one database system. The usefulness of this application is still under review and evaluation by the Waterworks. Further, the City has not yet invested in the implementation of a Laboratory Information Management System (LIMS). Instead, the laboratory uses a process by which samples are manually logged into the sample logbook, and then measurements are recorded in internal Excel spreadsheets or Microsoft Access databases. The measurements and analytical quality control results are verified by the laboratory director and, once approved, are uploaded into appropriate spreadsheets or databases that are accessible by other members of the Waterworks Water Production division.

Table C.8

Information and Water Quality Data Generation

Department	Type of Information	Collection	Storage
Water Treatment Plant Operations	Process Control Water Quality (e.g., organics, particles)	On-line, Grab	Electronic
	Operational Parameters (e.g., flow, chemical feed rates, tank levels)	On-line, Grab	Electronic
	SDWA Compliance Parameters	On-line, Grab	Electronic
	E-mails, Notes, Log	PC, Grab	Electronic, Hardcopy
	Reports	N/A	Electronic, Hardcopy
	SOPs	N/A	Electronic, Hardcopy

(continued)

Table C.8 (continued)

Information and Water Quality Data Generation

Department	Type of Information	Collection	Storage
Laboratory Services	Process Water Quality	Grab	Electronic
	SDWA & DEQ Compliance Parameters	Grab	Electronic, Hardcopy
	E-mails, Notes, Sample Log	PC, Phone	Electronic, Hardcopy
	Reports	N/A	Electronic, Hardcopy
Facilities and Watershed Management	Residuals Production	On-line	Electronic
	Land Application Rates	Grab	Electronic
	Soil Composition	Grab	Electronic
	Tree Growth	Grab	Electronic
	Vegetation Speciation	Grab	Electronic
	E-mails, Notes, Log	PC, Grab	Electronic, Hardcopy
	Reports	N/A	Electronic, Hardcopy
	SOPs	N/A	Electronic, Hardcopy

Numerous sources and users of information and data were identified. In addition to the plant and laboratory data, the City's billing system (PUBS), maintenance management program (Orion), GIS data and e-mail were identified as major data (or information) sources. Through the interview process a number of internal and external users of the generated data were identified (Table C.9).

Table C.10 summarizes the electronic data sources and the related internal data users. Communication of information generated by the Waterworks to external users is accomplished by the internal user being a liaison. No direct access to any internal data sources is permitted to external users.

Table C.9

Information and Data Users

Users within Public Works	External Users

❑ Treatment Operation
❑ Laboratory
❑ Facility Maintenance
❑ Engineering
❑ Distribution System Operations
❑ Facilities
❑ Customer Service/Public Outreach
❑ Director's Office
❑ GIS
	❑ Regulatory Agencies
	    ❑ Residuals Reports to DEQ
	    ❑ SDWA Compliance Reports to DHS
	❑ Other City Departments
	❑ General Public
	❑ City Council and Manager's Office
	❑ Consultants
	❑ County Agencies
	❑ Organizations/Businesses

Table C.10

Data Sources and Internal Users

Major Internal Users (current)	Major Sources of Electronic Data						
	SCADA	Laboratory Water Quality Data	Forestry & Residuals Data	Maintenance Management Program	Work Orders	Customer Complaint	GIS
Treatment Operation	X	X		X	X	X	X
Laboratory	X	X	X		X	X	X
Facility Maintenance	X	X	X	X	X		X
Engineering	X	X		X		X	X
Distribution System Operations		X			X	X	X
Facilities	X	X		X	X		X
Customer Service/Public Outreach		X				X	X
Director's Office	X	X				X	X
GIS		X				X	X

OBSERVATIONS

Our initial assessment in visiting Newport News is that it is an advanced utility relative to the handling and dissemination of water quality data. Newport News currently has a sophisticated information system but much of the information that is available is restricted to individual applications. In some cases the information is shared between different software programs, but little designed integration of information is currently occurring. While there are vehicles for sharing information there are still, in reality, a number of situations in which the data are held by a single entity and not normally accessed on-line except by specific arrangement. This is not to say that the data are not available, it is more a situation in which the system is not optimized to make the information available without specific arrangements made between groups. These arrangements may be a simple as notifying the person who needs the data where the data can be found.

The primary use of the SCADA system is for operations to monitor and control plant functions, but it does have some additional users. The maintenance staff utilizes the system to aid in trouble-shooting plant problems. The distribution staff uses the system for inline pressure data; however, they do not utilize the historical data at this time and stated the need for more training to maximize the utility of the data from the historical archive. The lab uses the system to read on-line instrument data. The lab had attempted to use the SCADA historical files for storing lab data but because of the need of manual entry this approach was not very successful. The historical SCADA data is utilized by operations for trending the data and daily reports. The SCADA data is combined with other data from the lab to be used on the monthly compliance reports. The SCADA historian is proprietary and has a built in report generator but no export utilities to other relational databases.

The Process Control Specialists have the technical expertise to handle most SCADA changes and upgrades in plant without outside consultants and keep a current on-going list of needed work items for repairs, enhancements and upgrades. The Process Control Specialists are aware of the limitations of the existing system and the abilities of other systems. For example, the historian does not have connectivity to any other systems. The ozone pilot project data is being collected via a third party software package that does allow the PLC data to be sent directly to an Access database.

The SCADA data seem to be well exercised within the Newport News Water Utility. In some instances, the system has been used for applications that are not really SCADA based (like storing water quality results) because the system works and is widely accessible.

Many of the laboratory measurements made by the operators are recorded on paper but are also manually entered into the SCADA database. A great deal of information about the operations of the plants is currently stored in the paper notebooks maintained by the operators.

This paper record is used to communicate between shifts and between different operators. This information would be better stored in an enterprise wide information system that could have input from SCADA, the laboratory (eventually a LIMS) and from manual entry done by operators and field personnel. This system could then be accessible for a wide range of users.

Lee Hall is in the process of building a brand new facility. Currently, there are almost no automated processes in the Lee Hall facility, with the majority of operations manually manipulated. Most of the information that is currently generated at Lee Hall is entered on a daily report and the daily report is also used as a source of input for the SCADA system. Any input to the SCADA system are done manually. A secretary enters daily information regarding water flows and rainfall manually. This manual entry has been ongoing since 1994. The majority of the data that Chris Harbin currently monitors is in the form of finished reports from each of the plants. He performs some aggregation of these results that may be recorded for long-term reference and passed along to the Utility management.

Unlike many other utilities, the plant operations and maintenance group is responsible not only for running the plant but also for delivering the finished waters to peoples' homes. Normally, this is the responsibility of the distribution system operations and maintenance groups. This is not the case in Newport News in which the scope of responsibilities of the Distribution Systems Operations group is strictly the maintenance for the infrastructure. Distributions and Maintenance is responsible for making sure that the distribution system is intact and functioning. They also have responsibility for maintaining the pumping stations. The plant Maintenance and Operations group is responsible for treating and distributing potable water to its customers over the pipes maintained by the Distribution System Operations and Maintenance. Plant maintenance personnel have access to a number of information systems that they use on a regular basis. In addition to the maintenance management, they utilize the SCADA system from which they get data but they also do some manual entry. They regularly need data and insights that in many other utilities are the preview of the Distribution System and Maintenance group. They are not in regular contact with the personnel in the field or with the customers. They require regular updates on complaints, scheduled maintenance and emergency repairs.

The plant Operations and Maintenance group has a number of additional responsibilities including:

- Watershed planning and management
- Residuals discharge into the watershed; specifically the dewatering of sludge and the spreading of the sludge on certain locations within the watershed.

As part of the watershed monitoring program, 12 wells are monitored quarterly at a minimum. Residuals of sludge spread on the watershed are analyzed twice monthly. The majority of the watershed monitoring and data processing is done manually. Results of most of the watershed work are not put into databases but are sometimes available via maps, spreadsheets, etc. Much of these data are mapped for visualization. The plant maintenance personnel are currently using GPS to properly locate stations in the watershed. Most of the available maps are currently accessible by any utility staff member that has access to ArcView over the Newport News Wide area network. Specific layers that are available include species and areas of application for sludge dispersal.

The plant managers and plant maintenance group felt that additional tools to support database creation, generation, and tracking would be desirable. They also indicated that they would like to have better access to lookup tables and material indices. Systems to help monitor and support decisions related to power management could help managers determine when to pump and when to be doing peak shaving, etc. These types of systems would likely pay for themselves quickly. They would like to have more information available in WaterNet and have that information be live rather than just data summaries. They would like to have more access to on-line budget information. They also would like to have systems that better manage and disseminate employee information.

Compliance reports that are generated include:

- The Virginia Department of Health - these reports are prepared by the laboratory, the operators and the plant superintendents
- The Department of Environmental Quality - these reports are prepared by the laboratory and the supervisor of the residuals branch.

Many of the reports that are generated extract the data from SCADA as well as from the water quality laboratory and bring the information together in either Excel spreadsheets or Word documents. Many of these aggregates are done on a monthly, quarterly, and annual basis.
The Busch brewery is a regular user of available data and is regularly looking for pH and other water quality data critical to their manufacturing process.

As indicated earlier, the Distribution Operations group is not responsible for supplying drinking water to the customers but is responsible for the infrastructure that allows the plant operations and maintenance personnel to deliver the water to the customers. This raises a potential problem in information flow since the distribution operations personnel are often the first to hear from customers when problems develop. When problems emerge, the Distributions Operation group is often sent out in the field to evaluate complaints and understand the cause of the problems. The Distribution Operations personnel indicated that often times they submit issues and samples to other branches within the Waterworks Department for analysis and resolution; however, they seldom receive any feedback once the problem has been diagnosed and resolved.

Most groups expressed satisfaction with the abilities of WIS to supply them with support and services. Backups are currently done on a regular basis on all central information systems and on all shared drives, but we heard of at least two instances in which data were lost within the utility within the last few weeks.

With the exception of the clearinghouse function performed by WIS there is no umbrella organization responsible for managing, disseminating and optimizing the use of information for the utility. Waterworks established a team to consider options for future development called Next Step. The team was organized in the spring of 2000 and their mission was to review and evaluate selected business workflow systems and processes that may be encumbering the organization and recommend what actions to take. By going through this exercise they will begin to develop the Waterworks Technology Roadmap. Their goal is to make Waterworks more cost effective and efficient, while continuing to provide a high level of service. This team is developing a strategic plan for the Waterworks. They are participating in the city's data processing team in the development of the strategic plan for the city.

Both the WIS Group and the GIS Group felt strongly that senior management has been a strong advocate for automated data processing within Newport News Water and has consistently given required support and resources to make this vision a reality. The WIS budget has been consistent at between $180,000 and $200,000. This is in addition to salaries and benefits. GIS has had a similar budget of approximately $150,000 per year. The two groups expressed concerns over the integration of the GIS capabilities with the WIS systems because the information systems have not and are not being linked to the GIS capabilities by design. Rather most of the data that is being integrated into GIS is either collected explicitly for GIS or is being extracted specially for inclusion in GIS coverages.

The GIS and WIS Groups have a good relationship, but there have been problems and issues. As observed with most situations in Newport News Waterworks, issues are routinely discussed in an open forum resulting in resolutions that are acceptable to all stakeholders. The WIS and GIS Groups are both concerned that the City has decided to go to mainframe information systems using DB2. Their concerns are over the conversion to a relational database that is not widely used within Waterworks currently. It will require extensive training and conversions of legacy databases into DB2. Because of focuses on Y2K, little upgrading has occurred within the City of Newport News during the last three years.

Most of the focus group participants felt that communications within the Drinking Water Utility were good and information-processing requirements were well understood. This positive environment allows the WIS and GIS personnel to serve other groups effectively, promoting communications and exchange of information. In general, the Waterworks believes in enabling people to participate in planning the future directions of the organization. Routinely, a large number of staff members are brought together for communications and discussions.

BARRIERS

The following barriers prevent the effective transfer of information data within the Newport News Water Utility:

- There is a lack of a centralized database for access to information. WaterNet tries to serve this role, but is a fairly static view of selected information.
- The SCADA system is proprietary in nature and currently does not have drivers to allow communication with standard databases. While the software vendor now

provides an option for ODBC connections, it will also require the Waterworks to upgrade their software version two releases to take advantage of the option.

- Several of the data platforms don't communicate well. A great deal of information and knowledge lies stored in a variety of systems, many with proprietary storage architectures. This prevents the definitions of relations to aggregate information in multiple systems.
- There are plans to standardize on DB2 as the underlying database for future information systems. This will introduce additional resource requirements for the conversion of legacy systems as well as training. This must be weighed against the cost of building bridges between the existing systems.
- There is a need for an information directory to inform users of the types of information that are available and to facilitate the location, access and use of these data. While a good deal of information is displayed in WaterNet, Waterworks staff do not necessarily know what resources are available beyond that in WaterNet.
- Some supervisory staff seldom uses email and rarely accesses information electronically. There are some cases of a lack of communication between divisions or branches.

RECOMMENDATIONS

In general, the Waterworks is fairly sophisticated in its use of information, but not without room for improvements. The following recommendations result from integration of all the information gathered during the focus group meetings:

- Develop a centralized warehouse of information. WaterNet is a good starting point and should be enhanced dramatically to include access to raw data in robust databases and summary information that is linked to the underlying data. This enhanced system should be accessed via the Intranet.
- Continue plans to get a LIMS installed and integrated into the enterprise wide system. The selection of the LIMS should be made based on an in-depth requirements analysis for the Laboratory. The LIMS should be implemented with

the goal of it being a laboratory support system. The LIMS should not be viewed as a repository of water quality data to be accessed by everyone in the organization. The data that are generated by the LIMS should be extracted, documented and formatted to be housed and distributed by the enterprise-wide system described above (WaterNet).

- A major priority for Newport News should be increasing the availability of water quality data and information. The enterprise wide system should be implemented before a LIMS is purchased. The Water Quality database should be considered a repository of all water quality data and information that is generated. The sources should include the SCADA, operator's data, LIMS and the distribution and maintenance group. This database should be implemented using relational database tools and should be accessible to all appropriate personnel within Waterworks and the city.
- The central water quality database should be used to generate standard report using automated routines. For example, the monthly report could be more fully automated with a common database. This would reduce the work hours needed to perform a scheduled repetitive task.
- A similar model should be considered for many operational data that have wide distribution. Intranet systems can be used to make the data available to a wide audience within the utility without making the data available to anyone outside of the utility. For example, the run hour times for plant equipment could be built into the maintenance software system to help determine scheduled maintenance activities on pumps and motors. Some of the automated reports could then be posted to the Intranet site for general access within the utility.
- Upgrade the SCADA system to provide ODBC connectivity to standard databases. There is also a need to expand training on the SCADA to secondary users.
- Historical data should be available in a relational database. This mobilization of the data will allow the pertinent SCADA data to be shared with other data processing and visualization tools, such as GIS. The SCADA upgrade described above would have to be implemented to allow this functionality.

- Begin standardizing naming and ID conventions for sampling locations, infrastructure, sample types and all codes. Names and codes are not universally standardized across the utility and therefore data sharing and access is further complicated.
- Formalize work order flow, especially for customer service responses. Some members of the organization want/need to be informed about the resolution to complaints and issues. Communication of the resolution of issues will give employees a sense of accomplishment and will also help with future problem resolution.
- System back-up procedures should be revisited department-wide. Standard operating procedures should be developed and implemented to ensure that data and information are not lost.
- Newport News is at a level of sophistication that suggests that Knowledge Management approaches could be utilized effectively. Specific areas that could be implemented successfully include records management, portals and data warehouses.
- Provide an ExtraNet capability for sensitive customers or special external users. The information supplied to managers and customers such as Anheuser-Busch from the SCADA system could be nearly real-time with a direct interface to the WaterNet.
- The plant complex is designed in a campus style and the operations group has great distances to cover during emergency situations. An alarm paging system that allows selected SCADA alarms to be displayed via an alphanumeric pager would be beneficial.

TUCSON WATER DEPARTMENT

UTILITY MANAGEMENT

Tucson Water is a department of the City of Tucson, Arizona. It serves both potable and non-potable water to a customer base of approximately 675,000 people, located in a 300 square mile area, within the Tucson, Arizona metropolitan area. The utility has a staff of more than 575 people. Figure C.3 shows the general areas within Tucson served by the Water Department.

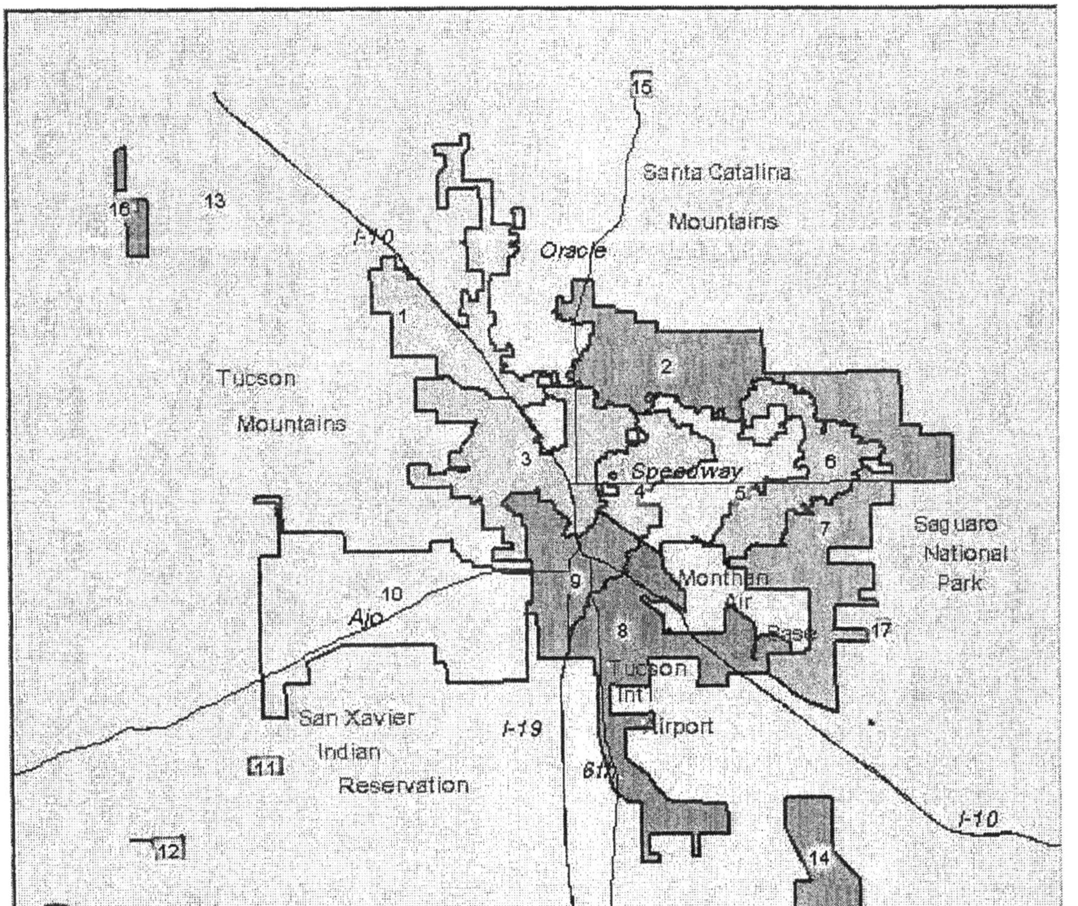

Figure C.3

Communities Served by Tucson Water

WATER TREATMENT PLANT OPERATIONS

All 37 plus billion gallons of water delivered by Tucson Water to its 600,000 customers each year has been groundwater pumped from area aquifers. In its 300-square-mile service area, Tucson Water operates 175 groundwater wells, 20 water reservoirs of 1 million gallons or more capacity, 3,500 miles of pipeline, and hundreds of booster stations and pressure control valves that give Tucson Water maximum flexibility in water delivery.

In the spring of 2001, Tucson Water began adding about 11 million gallons of water a day from their Clearwater Recharge Facility to their water distribution system. Colorado River water from the Central Arizona Project canal flows through a pipeline to specially constructed recharge basins. The water then soaks into the ground and, over time, reaches the water table where it slowly blends with the native groundwater. There are currently three recharge basins at Clearwater recharging about 5 billion gallons of Colorado River water annually. During the next few years, additional basins will be constructed to put utilize more of Tucson's allocation of Colorado River water to use.

PRELIMINARY DATA COLLECTION AND INTERVIEWS

Twenty questionnaires were completed and returned to the project team prior to their site visit. The interview and reporting process for the Tucson Water was conducted between May 7th and May 13th, 2000. Participants from the focus group meetings are listed in Table C.11.

OBSERVATIONS

Data Intensive Organization

The Tucson Water Department is a data intensive organization. Critical data and information are generated by a wide variety of departments. The information and data generated are critical to the daily operations and management of Tucson Water. Table C.12 summarizes some of the data generated by the different divisions.

Table C.11

Focus Groups Schedule

Group Name	Participants	Meeting Date
Senior Managers	• Marie Pearthree • Bruce Johnson • Mitch Basefsky	May 13, 1999
Division Administrators	• Sally Mapes - Water Quality • Larry Mulhearn (sitting in for Steve Pageau, Division Administrator of Planning & Engineering) • Ivey Schmitz - Business Services • Jim Broderick - Operations and Maintenance	May 8, 1999
Business Services/ Backflow Prevention	• Robin Williams - Backflow Prevention • Pam Humphrey - Human Resources • David Cormier – Finance • Gwen Goodman - Customer Service • Eric Unangst - Customer Service	May 12, 1999
Hydrology and OEM	• Marie Light • Bryn Enright • Terry Miley • Peter Chipello • Joe Babcock	May 10, 1999
Information Systems	• Mary Stewart • Nina Criss • Gail Dutcher • Randy Vanpool	May 7, 1999
Planning & Engineering	• Tom Victory • Rick Meyer • Sandy Elder • Larry Mulhern • Wayne Thureson	May 10, 1999
Water Quality	• Mike Dew – TWL • Tom Jefferson – TSD • Jean Melillo – CRS • Chris Pickett – TWL	May 12, 1999
Water Treatment	• Ted Bailey • Ray Wilson	May 11, 1999
Operations & Maintenance	• Lynn Baker – Preventive Maintenance/Maintenance Management • Cecilia Pena – Preventive Maintenance/Maintenance Management • Mike Ring – Water Treatment/Plant Maintenance Section • Keith Eagar – Water Treatment/Plant Maintenance Section. • Harold Maxwell – Equipment Maintenance • Rene Gomez – Plant Maintenance • Anna Marie Arenas – Service Maintenance	May 11, 1999

Table C.12

Data Produced by Functional Units

Functional Units	Data Produced
Senior Managers	Presentations, Reports to Mayor & Council, Directives to TW divisions, Public/Media Notices
Division Administrators	Presentations, Budget Reports, Requisitions, Mailings
Information Systems	Custom Queries, System Backups, Database Administration, Help Desk, Facilities/Phone Requests
Water Quality Lab	Water Quality Data, Compliance Reports
Planning & Engineering	Hydraulic Modeling Data, Land Use Plans, Capital Improvement Plans, Compliance Reports, Project Plans, Valve Maps, Wells Database, Permitting, Backflow Database
Hydrology	Compliance Reports, Presentation Materials, Hydrology/Wells Database
Water Treatment	Compliance Reports, Water Production Reports, Maintenance Management
Maintenance & Operations	SCADA, Compliance Reports, Maintenance Management, Water Supply System Summaries
Business Services	Materials Safety Datasheets, Accident Analysis, Vacancies, Training Data, Certification Data

Data and Information Focus is Division-Based

The majority of the data and information generated are focused within a single division. The stimulus for generating, maintaining and using the data is normally associated with a particular mission within the division that is creating, populating and using the data. Table C.13 lists examples of major databases and a listing of who generates them and who uses them. A reason that the database was developed is also presented.

Table C.13

Key Information Sources

Key Databases	Producers of Data	Consumers of Data	Use
SCADA	Operations	Maintenance & Operations, Hydrology, Planning & Engineering	Management of distribution of water, data source for management of routine maintenance
Water Quality	WQ Lab, Hydrology, OEM, Contract labs	Most sections, OEM	Tracks water quality for disinfection and compliance reporting
Wells	Maintenance & Operations, Hydrology, OEM	Operations, Hydrology, OEM, Water Quality	Compliance reporting, Maintenance
Training	Maintenance & Operations, Human Resources	City of Tucson	Provide proper training to staff, track certification renewals
Mailing Lists	All Sections	All Sections	
Regulatory Agencies, such as ADEQ		Water Quality, Hydrology, Planning & Engineering	Activity history, status of water sources and POEs, violations
County Records		Planning & Engineering, Human Resources, Business Services, Hydrology	Tracking land use
Maintenance Management	Maintenance & Operations, Water Treatment	Maintenance & Operations, Water Treatment	Maintenance scheduling and tracking, Compliance Reporting
COT Billing System		Most sections	Beside billing customers it is a key source of water consumption data and addressing for repairs and mailings
COT Project Management		Senior Managers, Division Administrators, Planning & Engineering, Business Services	
COT Budgets		Senior Managers	
COT Human Resources		Division Administrators, Business Services	Personnel Management, Compliance Reporting
COT Financial System		Business Services	
COT Payroll		Business Services	
COT Procurement		Division Administrators	
COT Business Licensing		Planning & Engineering	Backflow Prevention
Hydraulic Modeling	Planning & Engineering	Planning & Engineering, Maintenance & Operations	Capital improvement planning
Property Database	Business Services	Business Services	Management of department real estate
Presentations	All Sections	All Sections	Communications with the City as well as Public Relations
Material Safety Sheets		Business Services	Compliance Reporting
Backflow Database	Planning & Engineering	Planning & Engineering	Compliance Reporting

BARRIERS

Lack of Utility-Wide Understanding of Data Resources and Sources

A major barrier to the effective use and dissemination of data is related to the lack of knowledge of what information is available and how to gain access to the data. A directory of both information available and the mechanism for accessing the data is needed.

Majority of Data Exchange That Currently Occurs Is Customized and Based on Personal Exchanges

Most data exchange that occurs between divisions and departments occurs by individuals talking to one another and arranging for a customized exchange of information. While this is not an optimal means of data communications and sharing, it does have some positive features that are helpful. This approach fosters communication between groups and also discourages inappropriate uses of the data by requiring the user and the generator to talk to one another.

Data Entry Is Currently Difficult and Is Not Standardized

For many of the data management strategies currently being used within Tucson Water, data entry is difficult. Few if any of the current applications have standard or common methods for data entry. This situation results in users being required to learn and use a wide range of different interfaces. It also makes data entry prone to errors and inconsistencies, leading to problems in the quality assurance and documentation required for entered data.

Database and Information Redundancy

Due to the informal communication of data and personal exchange, many of the key databases are duplicated and maintained by a wide range of users. These databases are not regularly updated and in some cases modifications are made to the databases by different organizations with insecure and non-guaranteed feedback to the authoritative database. This is a dangerous situation that can lead to significant errors.

Geographic Information Systems and Data Visualization Is at a Minimum Desirable

There are many applications for mapping that could support different functional needs within Tucson Water. These include mapping of facilities, locating new developments, visualizing hydrodynamic processes and water quality. GIS is an important technology that can improve the efficiency and effectiveness of the Tucson Water department.

The City-wide Information Systems Are Difficult to Use and Are Not Optimal for Tucson Water

Tucson Water relies heavily on the City of Tucson information system and infrastructure. The Water Department is limited in the systems that it can use, the software it can purchase and the applications that can be run. There are benefits to being part of the citywide information architecture, but there are also serious limitations. The perception is that the limitations outweigh the benefits.

The Microsoft Office Suite Is an Effective Base System

The current standard for desktop operating systems and basic tools is Windows and the Microsoft Office suite. This is the standard and is working well. One issue raised was that when new versions come out they need to be adopted by the entire utility to ensure that files are compatible and accessible.

The LAN/WAN Is Effective Where It Is Available

Tucson Water's LAN/WAN, which is part of the City of Tucson Network, works very well where it is available. It gives access to shared drives, e-mail and the Internet. Some issues were raised over how the LAN/WAN was administered and difficulties in how shared files were accessed. As there is no single inventory of shared systems or users of those systems, notification is not possible when areas get reorganized or moved to different servers. As a security requirement, login accounts expire if unused for an extended period of time, creating difficulties for access, which might be needed only on an annual basis.

Citywide Systems Are Difficult to Access and Use

There are a number of citywide systems used as part of operating the utility. These include procurement (POTS), project management (PMATS), financial (AFIN), billing (Pueblo) and human resources (AHRS). These systems are often difficult to use, terminal based systems. Often systems such as the billing system require specialists such as in IS to retrieve the desired information.

The Pueblo Billing System Contains a Great Deal of Critical Information

The Pueblo billing information system contains a great deal of unique information that could be used very effectively by many people. Some of these data are currently extracted into a shadow file that is accessible but the data are limited. The vendor has been asked to include additional information but they have not been responsive. Since the Pueblo database and its interface are in a proprietary format, this work cannot be done by a third party.

Water Quality Data Are Critical to the Operation of the Tucson Water

Access to the Water Quality database has been difficult in the past. The new Water Quality database has resolved some of these issues. There are still some performance issues with the Water Quality database that are being addressed.

Inconsistencies

During the requirements analysis we focused on identifying discrepancies in the perception of information users. These are opinions that should be clear and agreed to by all users but currently are conflicting. Originally, we expected that these discrepancies had to have a single truth that would become clear in discussions. In some instances, seemingly diametrically opposed perceptions were both true based on the perspective of the person giving the input.

Many users indicated that access to data generated and maintained by other users and divisions was very difficult. The Information Systems (IS) section indicated that sharing of information between users was simple. All that the users needed to do was to put the data set on a shared drive and indicate to IS who should have access and then those people could access the data. Further discussion resulted in a clarification that the perception of IS was correct; however, once the data were available, privileges reverted and needed to be regularly updated. Problems still arise from reorganization of shared areas, migration of shared areas to new servers, and the expiration of logons to areas that are used perhaps once a year. There is no single inventory of systems and users of those systems.

The Wells database is a critical database for many users in the Tucson Water Department. The Wells database is a summary of the current and historical record of each well in the Tucson Water Department's system. It documents the characteristics about the well, its water quality and schedule of operations. Routinely, the authoritative database is maintained by Bryn Enright in the Hydrology division. This database is widely distributed and accessed by many individuals and departments. Because of its popularity, the database is often copied to individual workstations and accessed within a department. Rather than regularly reverting to a wells database in a shared directory, the copied databases are accessed and in some cases modified or extended. Part of the motivation of having the variety of wells databases stems from the variety of needs related to wells. Information retrieval about wells may relate to active wells, monitoring wells, private wells, or even just well sites that are property owned by Tucson Water and may not even have a well on them. Rather than using a common source, the different functionalities are handled by separate databases.

There are many human resource issues that need to be addressed in a database including training requirements and verifications, driver's licenses, safety and employment processing. Much of this information is required throughout the department but it is currently not easily accessed. Plans for distribution of both access and data entry have not been completed because of lack of resources.

The treatment plants currently use the Maximo system to track maintenance of reservoirs, boosters and wells. There are 22 computers which have Maximo installed with approximately 12 active users. Maximo is not used by all of the groups within Tucson water that could be using it. Tucson Water has also initiated a major review of requirements and available systems to choose and install a new maintenance management system.

REQUIREMENTS AND RECOMMENDATIONS

This section summarizes the major requirements expressed during the interviewing process and validated in the joint requirements planning. We will identify requirements that are both department-wide as well as those that are critical to a single division. The major identified requirements will be presented first with a brief summary then each requirement will be explored in greater depth with options, costs, and benefits.

The requirements that were identified fell into a number of major categories including:

- Directory of Resources
- Improved Communications
- Data Availability/Quality
- Data Processing Tools
- Administrative Systems

Each of these broad categories included a number of specific requirements that are presented and discussed below. As with any attempt at broad classification, some of the detailed requirements span a number of the broad topics. These broad topics are expressed here only to help organize the requirements. Details of the requirements can be found in Appendix C, which contains detailed notes of the requirements expressed during the focus group meetings.

Directory of Resources are databases, tables, spreadsheets or booklets that document what types of resources are available and how to get them. Specific related requirements include:

- Repository of presentation materials
- Index of information resources

Improved Communications include tools and mechanisms to facilitate the exchange of information within the Department, between the Department and the City, and between the Department and the public. Specific related requirements include:

- Enhanced web-based systems
- Tracking and sharing e-mail threads

Data Availability/Quality expresses the need for appropriate access to data and information of known quality that is maintained by one division for use by other divisions. At least one application, Back Flow Prevention, requires extensive information from a wide range of other databases and divisions. Specific requirements include:

- Distribution System Information Availability
- Extraction of Data from the Pueblo System
- Comprehensive Mailing List
- Access to Water Quality Data
- Access to Required Data for Backflow Prevention monitoring
- Access to Outside Databases including ADEQ, SDWIS and Pima County

Data Processing Tools include software and hardware tools that are used to access, process and disseminate the data to perform statistical analyses, hypothesis testing, visualization, interpretation, presentations and dissemination. Specific requirements expressed include:

- Geographic Information System
- Maintenance of Paper Documents and Files

Administrative Systems are information systems that help Tucson Water track requirements, paperwork, communications, reports and data. Specific administrative systems requirements include:

- Compliance Monitoring System
- Permit Tracking System
- Human Resources Information System
- Material, Data, Handling Sheets

The next sections address these requirements in detail. Many of the requirements described can be effectively addressed by a family of available technologies. Often the implementation of these technologies will address more than one requirement and in some instances, a particular requirement will need multiple technologies to adequately meet the needs. The requirements will be presented first and the technologies that can be used to address these needs will be presented afterwards.

- **Directory of Resources**

Tucson Water is an organization that is rich in data, information and knowledge. There are a number of people within the organization that have a strong understanding of the information that is available, how to get the information and what decisions can and should be made based on the available information. Unfortunately, there are no formal means to organize and inform the wider community of Tucson Water employees, other City department employees or the public as to what information exists, how to access it and who the information experts are that can help apply the data effectively. A clear requirement was articulated by the participants of this requirements analysis that organizing the available information and making it accessible to the appropriate people was critical.

This requirement addresses the need to know what information is available and who knows about the information. It does not address the actual on-line access to the information or the processing of the data. This step is just knowing that the resource exists and how to access it. In some instances the solutions proposed will also easily link the data consumer with the appropriate information. Some specific needs that were addressed for this type of requirement are covered in the next sections.

- **Repository of Presentation Materials and Reports**

Description - Tucson Water is in need of a management system to organize and locate presentation slides, reports, graphs and pictures. Presentations are made that include slides, pictures and concepts that could be reused effectively by others within the department for other presentations. This includes photographs, figures, pictures, graphs, and ideas. Often a number of people within the organization are making similar presentations to similar audiences. When critical photos, figures and graphics are needed, it is difficult to find these resources even though the potential user knows that they were developed in the past. Similarly, key reports and documents (e.g., those made to Mayor and Council, primary agencies, publications and public information releases) can not be easily located and referenced after they have been released. These documents may be critical and a precise record of what was said in them is important. Many of these reports have been stored in hard copy but not in a systematic way. Quick access and retrieval of these documents is important. The senior managers and the division administrators emphasized this requirement.

Options – Any successful solution to this requirement needs to allow different users to access a directory of available materials that can be readily searched in a variety of ways. The application needs to have flexibility for select members of the organization to update the presentations and to add new presentations. The solution needs to have version control so that the date of the presentation and any updates is known. The application has to facilitate the development and access to descriptions of each item (e.g. presentation, picture, slide, document, etc.). The capability to designate the appropriate security for accessing the materials is also necessary. The security capabilities should allow for the designation of levels of access so that where appropriate, different audiences can read and maintain the files. Not all users should have access for modifying or posting materials to the records management system.

Two overall options can be specified to meet this functional need. The first would be to organize presentations, images, documents, etc. under a set of shared directories in which different types of records could be maintained and made available to anyone with access to the Local Area Network. Access could be controlled by manipulating access rights to the directories. Appropriate documentation could be written for each file and maintained in the same directory. A number of different tools currently available could be used to embed pictures, graphs, slides, and presentations into existing files. While this approach could work, it would not allow easy searching and retrieval of objects. Searches would not work on embedded documents. This approach would depend solely on the user understanding adopted standards and maintaining the appropriate files and directories.

The alternative would be to implement a Records Management System application, which could effectively organize these resources. Records management software can be set up to work within the Microsoft Office suite to help users catalog files into an overall directory that can be shared by many users. The details of how the Records Management software works can be customized to meet the needs of Tucson Water.

Recommendations – This set of functional requirements can best be addressed by the implementation of a Records Management System for Tucson Water. This functionality will meet a number of the requirements expressed. There are a number of commercially available Records Management Systems that can be purchased and implemented on the City of Tucson Wide Area Network. These systems effectively develop and maintain a searchable index of any kind of files. The Records Management software allows users to search documents by type and for specific words in either the title or the file. A directory of the files that have been made available to the records management software is maintained. Specific documentation of the source, date, file type and an explanation of the contents of the file can be maintained. The Records Management Administrator can grant different levels of access to groups of people, allowing them to update, read, and delete files as appropriate. Any file type, including the types specified in this requirement, can be managed within the Records Management software.

Benefits - Implementation of records management software would facilitate the organization and dissemination of standard presentation materials. It would reduce the time and effort required to develop a presentation and would also allow Tucson Water to develop and maintain standard presentation materials. It is anticipated that this would result in increasing efficiency, simplifying preparation for presentation and reducing redundancies and inconsistencies in Tucson Water presentations. All tracking of files and documentation is done automatically and the user is prompted to enter the required information at the time of file creation or modification.

Costs - Records Management software is usually licensed on the basis of the number of copies of user software (often referred to as 'seats'), concurrent users and the type of server on which the main system is installed. The software is fairly expensive. It is likely that an installation for Tucson Water that would support the senior manager, division administrators and a limited number of other users (10 seats, 5 concurrent users) would cost in the range of $20,000-$30,000 to license the software. Costs for installation would depend on how much was done using Tucson Water Information System personnel versus outside contractors. Cost for customization and installation can be significant (range 100% - 300% of the cost of the license). Annual maintenance fees for these packages are usually about 15% of the licensing fee. For these tools to work effectively, a Records Management Administrator is required. This person would maintain the system, assign access and update records that have the highest security. The ideal candidate for a Records Management Administrator would have an information technology background but does not need to be a programmer or a database administrator.

- **Index of Information Resources**

Description - A need was identified for a comprehensive index of the available information resources that could be accessed by users within Tucson Water. This functionality would allow any data consumer related to Tucson Water to identify critical data resources. In the past, Tucson water has had a directory of information and resources that was printed and accessible in a hard copy. This directory has not been updated recently and nothing has replaced it. An effective solution to this requirement would include sufficient information to identify where the information resided, to whom it is accessible and how to access it. The directory would also contain documentation indicating who collected the data, why it was collected, how it was collected and analyzed, as well as a link to related files and a contact name to get more information. Basically, this would be a data directory and data catalogue that organizes available information and expertise. In some cases, the access might include a contact person and in other cases hotlinks to the actual data which are available.

Options - These requirements can be effectively addressed using either a high-tech or low-tech approach. The low-tech approach would involve compiling a booklet that documents the critical data resources and the key information resource people within Tucson Water. This booklet could be distributed as hard copy for people to use at their desks or via electronic files at their workstation over the Local Area Network. The high-tech approach would be to develop a database that was accessible as a web-enabled information system. The directory of information resources could contain not only databases but also technical experts. The databases could be easily cross referenced to the experts that are responsible for or knowledgeable about each database. The database could have a set of standard reports that could be used to generate a hard copy booklet that would be the same as the one described above for the low-tech solution.

Recommendations - The most efficient and enabling solution is to build a database and resource directory that is web-enabled and made available through a standard Web browser interface. The low-tech solution outlined above will require the same data collection that would meet the needs of the high-tech solution. The high-tech solution will codify the information and make it easily accessible to a wide range of potential users. The initial costs of a low-tech approach will be significantly lower than for a high-tech approach. However, maintenance of the low-tech approach is likely to be more difficult than the high-tech approach. The high-tech approach will also enable the generation of a hard copy document that would meet the needs of those users needing the information when able to access a computer. The high-tech approach will allow a wider range of users to access the data and will also facilitate the maintenance and flexible access to the required information. The database approach will allow customized queries and access to the information and could also facilitate hotlinks from the directory to the actual databases. Custom queries could also be used to generate useful information for dissemination of data to other city departments and the public.

A number of models for data directories can be evaluated and considered as models for implementing an information directory for Tucson Water. It is unlikely that any existing directory will meet the needs of Tucson Water. The database can be implemented as an Oracle database with a web interface. The interface should be developed to control data maintenance, access and application administration.

Benefits – Regardless of the approach taken, the development of a directory of information resources will improve efficiencies by informing personnel of the information resources that are available and the individuals within the department that can support the access and correct use of the data. The directory will reduce the amount of time individuals spend trying to locate pertinent information. The directory also assists new or relocated personnel that do not have the experience of tracking down the information. The directory will also reduce the redundancy of data collection, database management and quality assurance. Coupled with the Records Management System, the directory could maximize the use of data to support decision-making, operations and water quality.

The implementation of the recommended solution would make the directory available as an online resource and would facilitate control of access to the data. Direct links could be established between the directory and databases, simplifying access to the data. The on-line links could also be used to control how different users view the database by customizing the information shown and the methods of querying the database.

Costs – Tucson Water already has access to the Oracle Database Management System as part of the City of Tucson enterprise wide Oracle license. Any directory of databases that is developed will require design and implementation by a combination of a database administrator (or designer) and a database programmer. It is likely to require a database design effort and a Rapid Application Development process. It is possible to implement a system like this with a few months of effort. Tucson Water has the required in-house expertise but might consider contracting out this effort. In order for the directory to work effectively, it must be maintained properly. Periodically this database will require a quality control check to ensure that the data is current. Any links in the database will need to be tested on a regular basis (this can be automated as part of the system development). If this system is not maintained properly, then it will not be useful. It will require at least the part-time attention of someone who sees this as a priority. Recommendations for who this should be are presented in the Information Plan below.

- **Improved Communications**

Users indicated that one of the major improvements required within Tucson Water was an improvement in communications within the Utility and the City. Many possible approaches were discussed. We focus here on information technology approaches that could be used to improve communications and convert communications into useful, accessible information. These tools and mechanisms need to facilitate the exchange of information within the Department, between the Department and the City and between the Department and the public.

- **Enhanced Web-Based Systems**

Description - Tucson Water has a public web site (http://www.ci.tucson.az.us/water/) that is part of the Tucson City web site. The official Webmaster that controls content and maintenance of the system is part of the City of Tucson Information Technology Group. Tucson Water has a local web master who is a member of the Information Services section. The City has robust requirements and severely restricts what can be deployed on the City web site. The current web site is targeted at the public although it contains information of use to Tucson Water employees and other interested City employees and officials. The Senior Managers expressed a need to be able to post information on the existing site more conveniently.

A requirement for organizing and disseminating data within the Tucson Water Department and other associated departments was also expressed. This functionality would not be targeted at the public but rather aimed to be used internally. This functionality should also include the links to all other pertinent information, systems and resources available within Tucson Water.

This requirement was identified by the Division Administrators and Planning & Engineering. During the Joint Requirements Analysis this need was supported by almost all other groups.

Options - It is important for Tucson Water to continue to participate in the City-wide web site and it is appropriate that materials targeted for public consumption adhere to rigid standards established by the city. There is also an opportunity to use a web site for internal communications and for organizing information resources. This type of implementation is often referred to as an Intranet. The contents are accessible only to employees of Tucson Water and to other individuals who are explicitly approved. This approach could be used to make information available internally and develop the focal point of information exchange among Tucson water divisions and employees.

The technical options for implementing an Intranet site are numerous. Any implementation will require someone to oversee the site. Some standards will need to be adopted and maintained. It would not be advisable to make the site completely accessible to any staff members to update. Those wishing to post materials could do so through contact with an authorized person within their section. At a higher level of sophistication and cost, an intranet site could be implemented that would include a bulletin board on which employees could post information, links, data, etc. of interest to other Tucson Water personnel. A bulletin board would allow anyone with access privileges to post and retrieve messages. At a still more sophisticated level, Tucson Water could purchase portal software that would allow a webmaster to designate what view is presented to each user (or group of users) and what privileges that user has for each component of the web site. Portals can also be set up to allow individual users to determine the content and the format of what they access over the Internet.

Recommendations - Portals are very popular now but seem to us to be excessive for the perceived need within Tucson Water. Therefore, we recommend that portal implementation not be undertaken at this time but rather an intranet with bulletin board capabilities be considered. The Intranet application that is implemented should be designed to be the hub of the information organization discussed in this document. A well-maintained Intranet can be used to communicate many ideas for Tucson Water personnel. The Intranet site can also be used to distribute basic forms, memos and other resources. Tucson Water should continuously monitor this site to determine if the staff is utilizing it. If the Intranet site becomes a popular and effective tool, then the City should be encouraged to purchase and implement portal software.

Benefits - Implementation of an Intranet would meet a number of functional requirements by using a simple mode of access that is readily available to many potential groups. The use of a standard browser interface would allow even unsophisticated users to begin accessing the information base. The Intranet site could be used to organize and distribute forms and information. Users could post data assets for others to access and use. This would present management with a way to communicate effectively across divisions. Use of the Intranet site could also facilitate rapid dissemination of information when it is required.

Costs - Implementation of the recommended solution would not require additional software or hardware. However, an employee would have to be responsible for this system. The initial development could easily be done via contract but once it was running, it would require maintenance, updating and content development by Tucson Water personnel. This responsibility could be dispatched by the same person who is maintaining the Records Management System. Implementing a portal system would be very expensive for Tucson Water, making it an ill-advised approach unless it is adopted by the City of Tucson. For an entire utility with over 500 employees, it is likely that a license for portal software would cost between $50,000 and $100,000. Committed staff would be required for both system and content management.

For this approach to be useful, all members of the Tucson Water Department would need to have access to the Internet from their desktop.

- **Tracking and Sharing E-mail**

Description - Significant management decisions, discussions and reports are conducted and distributed via e-mail. Often times these communications are critical and the contents need to be shared and maintained as an enterprise information resource. For some of these communications, access needs to be carefully controlled. The employees that need access require quick availability of the entire communication thread (the original message and all follow up messages related to the original). These critical e-mails need to be searchable by subject and text to allow users to find appropriate e-mails that may span many years and subjects.

An effective implementation of this functionality will allow users to easily catalog their e-mails and specify who can see the results. The users will also be able to specify how long the e-mails should be maintained on line and what to do with them after the cut off date is reached. These functions could be best served by a system that can archive and restore old e-mail threads without great difficulty. This requirement was articulated by the Senior Managers and the Division Administrators.

Options – A wide range of options for meeting this need can be listed. To limit the number of options discussed here, we will focus on three levels of technology and recommend a solution that optimizes the utility of this type of functionality.

Many of these critical e-mails are currently stored as hard copy and filed in traditional cabinet filing systems. This makes retrieval, archiving and searching difficult and tedious. It also makes it difficult to ensure that appropriate parties have access to the e-mail threads and share common versions. While this paper approach is easy to implement, it does not meet the needs of searching, archiving and controlling access.

The next level of technology that could be applied to meet this requirement involves the use of existing e-mail and file management capabilities. Clearly e-mails can be forwarded with the entire history of the thread using standard e-mail software. These e-mails can either be stored in user or shared folders or can be saved to shared directories. This approach does not require new approaches, software or culture. It can be used effectively in small organizations in which everyone understands the file structure, the protocols for distributing e-mails and the mailing lists that need to be included in each mailing.

The third option is to use a Records Management System as described above. Records management would allow the organization, management and dissemination of e-mail and associated hard copy communications using tools that can be easily built into most desktop e-mail systems.

Recommendations – Records Management software would address a number of requirements expressed by Tucson Water. Records management integrated into GroupWise would allow users to instantly decide which e-mail threads needed to be made available to particular co-workers. Groups could be set up so that important messages could be managed and disseminated at the same time that they are read. The effort involved would be minimal and it could easily be done by managers. Storage time and automatic archiving could be specified. All employees with appropriate file privileges would be able to retrieve individual e-mail messages or entire threads.

Benefits – Implementation of a Records Management System would result in Tucson Water being able to manage its information resources more effectively. It would reduce redundancy and ensure that users are viewing the correct version of files regardless of the format. The application of Records Management to e-mails would enhance a favored form of communications into a chronicle of discussions, decisions, authorization and directives. E-mails would become an information resource for groups of workers and would improve efficiencies in managing, storing and retrieving critical documents and discussions.

In addition, the implementation of records management could facilitate using e-mail as an effective and official means of communicating with clients, developers, contractors, regulators and other city departments. All correspondence could be indexed, organized and stored with controlled access, searching capabilities and planned archiving. This could result in a positive change in the culture of how business is conducted by the Tucson Water Department.

Costs - The costs of the records management software and the associated personnel are listed above in on developing a repository of reports and presentation materials. Since this same functionality would supply a number of functional requirements, the effective cost of acquiring the software, installation and maintenance decreases when normalized to the functionality achieved.

If a Records Management Systems is implemented, it could meet this functionality with off the shelf systems. Records Management software can easily be incorporated within most existing mail systems and e-mails. Threads and even hard copy mail messages can be easily cataloged and associated with one another.

Data Availability/Quality

The need for access to authoritative databases that are designed and maintained for wide use within Tucson Water was clearly emphasized by all participants in the requirements analysis. Details of the specific requirements for each division are documented in Appendix C. Many critical data sources are either accessed through personal contact or by making copies of existing databases and then keeping a copy, which may or may not be updated on a regular basis. It is possible for multiple employees within Tucson Water to be basing decisions on different copies of the same database. Sometimes the quality of the data is not clearly documented within the database (including last update, version, etc.). The following information that could be integrated into a single large central database or built as separate databases was addressed explicitly during the requirements analysis as major requirements of broad interest and importance within Tucson Water. Many of these information requirements overlap. Needs for the data vary based on the decisions and actions involved. In some cases, a central database is of critical importance to a number of divisions and sometimes to other departments within the City of Tucson. In other cases, decisions and actions need to be based on data extracted from a wide range of databases. In these instances, the ability to correctly cross reference data is of critical importance to ensure that proper conclusions are reached.

The following summaries represent the major information needs that should either be made available from centralized databases or that require integration from a number of different databases. The priority for these is discussed below in the Information Management Plan.

- **Wells Database** – This includes information about each well that is a part of the Tucson Water Department system. The database records specific information about each well and its operational data including well construction, well maintenance, pumping rates, hours of operation, land ownership, land use, permitting and regulatory compliance reporting on wells. There is widespread interest in this database. It is a critical component to the implementation of any compliance monitoring system since well operation will dictate some important criteria for monitoring. The existing wells database is a set of databases with variations in focus and is used by many divisions and individuals and should be considered a very high priority for implementation. This requirement was identified by the Senior Managers, Division Administrators, Water Quality, Hydrology, Information Services, Planning & Engineering, Operations & Maintenance and Water Treatment. During the Joint Requirements Planning meeting, the importance of this database was validated by the entire group.
- **Data from the Pueblo Billing System** – The City of Tucson maintains a centralized Billing system called the Pueblo Billing System. It contains core information that is

of importance to most divisions within Tucson Water. This database seems to be the best maintained since it is used to bill customers. It is a major source of mailing information, changes in use, maintenance requirements, water usage, estimates of growth and other fundamental requirements of Tucson Water. This requirement was articulated by the Distribution Design and Plant Design Units, Operations & Maintenance, Business Services, Backflow Prevention, Planning & Engineering, Information Services, and the Division Administrators. Currently most of the access to information from the billing system is facilitated through a customized extraction from the central Pueblo system (referred to as the Shadow File by most Tucson Water personnel). This extraction is done and maintained by the IS group. The current shadow file needs to have additional information extracted to meet growing needs. The requirements for this enhancement have been developed through support of the IS group. Unfortunately, the modification of the shadow file extraction cannot be done by anyone other than the vendor, and the vendor has been unresponsive in implementing the request.

- **Mailing List** – There are currently a number of mailing lists that are maintained within the Department. Some of these lists overlap while others target specific constituencies and stakeholders. The list that is required is a database that merges all of the existing mailing lists and allows coding of the addresses to determine the type of contact listed. The mailing list needs to have permanent links to other databases (like the Pueblo shadow file) for automatic updating. Similarly, an interface is needed to allow updating, including addition of new addresses, new categories for existing contacts, deletion of old contacts, changes of names, etc. This requirement was specifically identified by the Planning & Engineering and the Director's Office/Senior Managers groups. During the Joint Requirements Planning, all groups agreed that this is an important database that should be implemented.
- **Water Quality Data** - This data originates from the Tucson Water Quality Laboratory, contract laboratories and field samplers. The data is currently contained in the Water Quality Database that came on line in December 1999. This requirement was identified by the Senior Managers, the Division Administrators, Water Quality, Hydrology, Information Services, Planning & Engineering,

Operations & Maintenance and Water Treatment. It is important to note that to some of these groups, particularly the Senior Managers and the Division Administrators, the individual measurements are not as useful as summary information. The current water quality information system does not present summarized water quality data. It does make the required data accessible for summarization using other tools like statistical software, GIS and spreadsheets.

- **Supervisory Control and Data Acquisition (SCADA) Data** – Tucson Water's three SCADA systems gather a significant amount of data. **Valmet System** manages the wells. The **Formey System** manages the Hayden-Udall plant and the **PLC System** controls the TARP plant. The data collected includes wells in service, flow rates, basic water quality data, reservoir levels and other data that could be useful to many Tucson Water divisions. The information collected is available but only through specific requests to the Operations & Maintenance Division. All extractions need to be customized, but once customized they can be delivered on a regular basis. Most extractions require some aggregation of the results because of the extensive and detailed nature of the data collect by the SCADA systems. This requirement was identified by Water Quality, Planning & Engineering, Operations & Maintenance and Water Treatment.
- **Backflow Prevention Monitoring** – The requirements of the Backflow Prevention group are illustrative of the complexity that can be required within Tucson Water for a variety of data types in order to perform a critical task for the Water Department. Near real time communication of changes in the ownership and use of facilities is critical. The Backflow Prevention group listed the following information requirements to support their effective prevention of backflow:

  - Pueblo Billing System
  - Customer Changes - These changes are reported to the Backflow Section by e-mail from the Telephone Section Staff, Front Counter Staff and Business Office Clerical Staff
  - Meter Changes - This information is provided by the Operations Division's Meter Shop

- Business License and Tax information - If the water bill is paid by a Property Management Company, the business may change from a low hazard Insurance Office that does not warrant the installation of a backflow assembly, to a doctor's office that would be considered a potentially high hazard account
- New Water Service additions - This information is provided by the Planning & Engineering's New Development staff at the time of a new water meter purchase
- Property records available electronically through the Assessor's Office are utilized frequently

Effective and accurate communication of this information can facilitate the efficient processing of over 25,000 Backflow Prevention accounts. Most of the availability and communication of these data is currently completed through arranged communication between divisions. These data could be organized into databases so that the notification is automatic and the data can be accessed repeatedly as required.

- **Outside Databases including ADEQ, SDWIS and Pima County** – Users within Tucson Water need regular and easy access to these outside databases in order to determine what the State, County and Federal Government have on record for the Tucson Water Department. Critical access includes: Waiver maps showing contamination sources in vicinity of wells (submitted to ADEQ) – prepared by Pima Association of Governments (PAG), Pima County Base Maps, Recorded Deeds, and Recorded Easements. Access to State and Federal regulatory agencies could be used to monitor reporting of Tucson Water compliance variances, ensuring that incorrect entries in EPA and ADEQ databases are flagged and corrected.

Options – To facilitate the sharing of these and other critical databases that are needed by multiple divisions within Tucson Water, new relational databases should be implemented. The implementation should be performed as an integrated system that will maximize the ability to combine data from different databases. This capability will allow users to combine data in many different ways to address the issues listed above and to address other needs that will arise in the future. Each database listed here needs to be implemented following a detailed requirements analysis that is beyond the scope of this project. The requirements analysis needs to document exactly what information is required, sources of the data, data entry requirements, quality assurance needed, levels of access, required reporting and database administration requirements. There are not many broad options regarding how these data can be managed and mobilized to support efficient management and decision support within Tucson Water. The status quo of informal communications can continue but it will result in information use not being optimized in Tucson Water.

Recommendations - We can list many other options, but in our opinion the requirements are best served by the implementation of a relational database system that is made available through a client/server-based system. Wherever appropriate, access to the database should be designed and implemented through web browser based interfaces. Access to outside databases should be done using interfaces developed by the suppliers of the data in the short term. Tucson Water should adopt standards established by Pima County, ADEQ and EPA wherever possible to promote data sharing and eventual seamless access of outside data through web-based applications. Much of the available data will be able to be accessed over the Internet in the near future using interfaces developed by Tucson Water that access both data specific to Tucson Water and other sources. It is important that most of these databases be designed and implemented with consideration of both the Maintenance Management System being considered and the GIS system implementation which are underway.

Benefits – Relational databases can be developed to address the specific needs that have been identified and would be further refined in focused requirements analysis. Well-designed databases will also provide the ability to address information needs that have not yet been identified but are likely to emerge in time. Adoption of standards within Tucson Water will improve efficiency, reduce redundancy and enable easy sharing of information between groups. Very significant improvements in efficiency and effective decision-making could be realized, enabling restructuring of personnel and departments.

Costs – Design, implementation and maintenance of good relational databases is a costly process. The basic tools required to implement these databases are already available within Tucson Water. The Tucson Local Area Network, servers and desktop systems are adequate to begin implementing the systems listed above. The City of Tucson software resources, including an enterprise license for Oracle, multiple copies of Microsoft Access and reporting tools (Crystal Reports) are already in place, negating a need for extensive investment in software. Major costs related to this approach would be realized in personnel, training and changing the culture within Tucson Water. The implementation of all of the systems listed above would probably take a few years of effort on the part of two to four full-time database professionals. Users would need to be trained in the new systems and professional database administration and support would be required to maintain the databases. While critical skills are available within the current IS staff, additional personnel would be needed to undertake all of these projects. Subject expertise to support database design, implementation, testing, and maintenance would be required. This is a direction that many large businesses are now taking, resulting in significant improvements in efficiencies. However, implementing a database system of this type will take a significant commitment by the entire utility to change the way business is done.

Data Processing Tools

Data processing tools are defined here as major information systems that are used to process existing information. Few distinctive needs for new data processing tools were identified. There are a number of software and hardware tools that are used to access, process and disseminate the data to perform statistical analyses, hypothesis testing, visualization, interpretation, presentations and dissemination.

- **Geographic Information Systems**

We touch on Geographic Information Systems here but refer the reader to the Tucson Water GIS strategic plan, which was being developed concurrently with this requirements analysis. Further details of that effort can be found on the web site http://www.terrasw.com/terrasw/tucwater/statplan/index.htmlx.

Description - Many people involved in the requirements analysis indicated that they felt that maps were critical and the ability to visualize a wide range of data would be helpful throughout the utility. There are a number of significant efforts underway within Tucson Water to use GIS and to enhance the utility-wide access to GIS tools and data.

Geographic Information Systems are being utilized effectively by many large water utilities comparable in size to Tucson Water. While Tucson Water has some instances of utilizing GIS in Planning & Engineering and Hydrology, geo-spatial information is in general underutilized at Tucson Water. A primary reason that GIS technology and geo-spatial information have not been more effectively utilized is the lack of a digital version of the Valve Maps/Distribution System. If the distribution system was available as a GIS database and linked to the Pueblo Billing System (for customer information), customer service, operations & maintenance and others, it would be in a position to provide improved information inter-departmentally and to the public.

Currently 400 out of 1,400 quarter section valve maps have been converted by internal staff using AutoCAD. In 1999, Tucson Water conducted a comprehensive GIS strategic plan with the assistance of GIS consultants (GSM and TerraSW) that were very familiar with the Department. The primary recommendation of the study was that the valve maps be converted into a GIS. The study cited the fact that over 3,000 person hours are spent each year updating and distributing the valve map books within the department – primarily as a manual process. Having a digital version of the valve maps and map books that could be brought to the field by Operations and Maintenance staff, and as a map view for the customer service representatives would present cost savings and operational efficiencies not currently attained at the department. The GIS strategic plan provides a detailed cost benefits analysis of implementing GIS in the department and should be referred to for a complete understanding of these advantages.

While the conversion of the Valve maps is a key first step to implementing GIS at Tucson Water, effective access to the Pueblo Billing system, staffing, support, training, and application development are other areas that will need to be addressed to optimally utilize GIS. Again the GIS strategic plan provides a good road map for addressing these areas.

Options – Ideally, Tucson Water would embark on a full scale conversion of the valve maps into a GIS database to have a complete vector-based representation of the distribution system. This fully converted vector data would support all potential applications within the department; mapping, work order maintenance, hydraulic modeling, customer service, planning, new area development, and several other areas. If the cost and timeframe present barriers to begin this effort, then an option is to scan the valve maps and convert just the distribution system to vector linework. This, along with a digital streets file and geocoded customer addresses, would allow for many of the applications noted above to be initiated in a shorter timeframe and at less cost. As-built drawings could also be scanned and geo-referenced to map-based features (customer locations, street intersections). While scanned valve maps would not provide all the potential uses of GIS in the department, it would provide a means for Customer Service, and Maintenance and Operations to locate customers quickly and with visual map based reference.

The GIS Strategic Plan presents the possible application cost and implementation time. These were referred to as Quick Start Applications in the Strategic Plan. These applications present other opportunities for shorter-term applications that can be supported more expediently than others and should be reviewed and considered.

Recommendations – Effective expansion of the use of GIS in Tucson Water will depend on three major components: 1) availability of good core data coverages, 2) training of the people who need to use Geographic Information Systems, and 3) the availability of support for accessing, using and analyzing spatial data. Emerging capabilities for the ESRI tools that are standard at Tucson Water will allow a technical staff to develop specific applications that can be used by a wide range of users, including public contact specialists, secretaries and maintenance personnel. Also, these applications can be web-enabled to allow users with web browsers to use sophisticated mapping tools.

TPMC recommends that the Department review the costs and timeframes to scan the 1,400 valve maps and convert to vectors the distribution system that have not been converted to AutoCAD format. Finding a means to attain a complete digital version of the valve maps that can support future maintenance management applications (currently under development), customer service, and operations will provide Tucson Water new efficiencies in inter-departmental communications and customer service.

As noted in the GIS Strategic Plan, effective access and information exchange with the Pueblo Billing System is the other key component to geo-enabling the department. It is essential that direct access to the customer data be provided from GIS based applications. Otherwise redundant storage of "shadow files" that become quickly dated will continue to present inefficiencies in managing customer and facility data.

Benefits - By moving forward on geo-enabling the maintenance, operations and other staff that maintain the distribution system and interface with customers, the Department will provide better customer service and trouble response. Having a digital version of the valve maps will cut down on the time, cost and inefficiencies of manually editing and duplicating maps for distribution within the department. Staff time spent locating maps or finding most current information will be decreased.

Costs - The costs to fully address the conversion of valve maps, training, GIS application development and implementation an effective billing system linkage was determined to be $4.6 M in the Strategic GIS Plan with a cost saving of $5.3 M as a conservative estimate. The Valve map conversion itself was estimated to be $1M over a two-year initial period with an additional $1M estimated to maintain the developed maps. The costs for pursuing the incremental approach of scanning the valve maps with limited conversion of the distribution system would need to be provided by a Data Conversion vendor.

- **Maintenance of Paper Documents and Files**

Description - Many documents developed by Tucson Water, land developers, engineers and other departments are routinely developed on hard copy documents. These include blue prints, permits, work orders, well construction and historical documents that are not available electronically. The requirement expressed included the ability to scan these document, store them electronically, develop a searchable index that includes a description of the document and then the ability to retrieve the image and reproduce the hard copy. This requirement was identified by Hydrology and Senior Managers.

Options - This requirement can be met through the use of either document management or records management applications. Document management is optimized for scanning and handling of both electronic and scanned documents. Records management can handle this requirement but it is not optimized for scanning, handling and management of scanned documents. Wherever possible, Tucson Water should begin moving towards submission of documents in electronic form to eliminate the need to scan documents and store images. Electronic documents can be indexed in records management software and can therefore be organized and retrieved along with scanned documents.

Recommendations - Since records management would meet so many other requirements expressed during the requirements analysis and can be applied effectively to manage scanned documents, it should be applied to this need as well. In order to determine a strategy for scanning existing documents, a detailed requirements analysis is required to understand the scope of the existing documents and what format they currently exist in. For example, scanning of existing documents that are in an unbound form would be handled differently than scanning bound documents.

Benefits - Use of a common records management application for a wide variety of object types would result in the development of a comprehensive index of information resources. The same index would include slides, images, presentations, reports, scanned documents and electronic documents.

Costs - Depending on the types of documents that would be scanned and the required resolution, the cost of scanning existing documents can be costly. The details of the potential cost depend on how much of the scanning will be done in-house, the number of documents that are anticipated per unit time, the format of the documents, and output requirements.

Administrative Support Requirements

These requirements are grouped together because they represent tracking of critical information, report submissions, regulations and safety information. The Human Resource requirement already has a system that has been purchased for addressing the need. The implementation has not been successful and needs to be revisited. The other requirements are either new altogether or have been handled using paper-based systems.

- **Compliance Monitoring System**

Description - Compliance with regulations is a complicated task given the number of regulations that need to be considered and the complexity of the timing identifying which resources require monitoring and reporting. The combination of these considerations results in a complex situation. Sometimes monitoring requirements are missed due to the complexity of the operations within Tucson Water with so many points of entry and wells to monitor and report. Compliance reports are currently submitted to ADEQ, EPA, ADHS, ADWR, and Pima County.

The required system would record and remind personnel of known monitoring requirements and would alert utility personnel of potential compliance monitoring for the wells and points of entry (POE) to the distribution system. To be effective this system would need to interact with the Water Quality Database and with a comprehensive Wells Database. The wells database needs to include information about the in-service dates and when a well went out-of-service. These dates need to be considered relative to monitoring requirements for each regulation. The system would need to analyze and aggregate data from these other databases along with information unique to the compliance monitoring system to evaluate and summarize at a minimum:

- Completed monitoring by POE
- Whether the initial compliance monitoring year has been completed
- Whether there have been any detected contaminants
- Whether the source has received reduced monitoring or monitoring waivers
- Whether there are additional discretionary monitoring

This requirement was identified by Water Quality and the Senior Managers.

Options – We have investigated the availability of a Compliance Monitoring System that would meet Tucson Water's requirements and none were found that meet these needs.

Interaction with the Water Quality Database and the Wells Database would result in any commercial system requiring some level of customization. Many states are working on systems to monitor compliance of the utilities that they regulate, but none are yet available. The other viable option is to begin implementing a customized Compliance Monitoring System. It is important to note that in order to undertake this implementation, a comprehensive Wells database would need to be implemented as a prerequisite.

Recommendations - The best solution for Tucson Water would be to design, implement and deploy a Compliance Monitoring System. This implementation should fold in the implementation of a Wells Database. The Wells Database is a requirement that is needed by many groups within Tucson Water for a variety of reasons. The implementation of the Wells Database and Compliance Monitoring System could be undertaken concurrently but should be done by an integrated team. The implementation of the Wells Database needs to be carefully interfaced with the Maintenance Management System evaluation and implementation since the Wells Database should be closely related to, if not an integral part of, the Maintenance Management System.

Tucson Water could consider teaming with a private contractor to do this with the goal of eventually commercializing and selling the system to other utilities.

Benefits - Implementation of a Compliance Monitoring System would help ensure that all compliance requirements are organized and accessible by the appropriate personnel within Tucson Water. A Compliance Monitoring System would also alert Tucson Water personnel of potential requirements for monitoring POEs that might otherwise be missed. The system could result in reduced penalties and monitoring imposed by the regulating authorities.

Costs - Implementation of a system of this kind will require a significant effort on the part of both database developers and technical personnel. The technical personnel will need to meet with the database team to clearly identify and verify the documented requirements. The entire process will likely cost between $70,000 and $100,000. These estimates do not include the design and implementation of the Wells Database.

- **Permit Tracking System**

Description - Tucson water applies for permits that are critical to operations. These permit requests are submitted to regulatory agencies and include requests for well development, wetland recharge and regulatory variances. Tracking the permits is critical to knowing the status of the requests, tracking communications, responding to objections and monitoring conditions imposed. This requirement was identified by the Senior Managers and Hydrology.

Options - Permit Tracking Systems have been implemented and used by states and the EPA for many years and many systems are available commercially. Most of these systems are built on a basic design that includes database structures, entry screens and reports. In most cases, these systems need to be customized to meet the specific needs of the organization. It is very possible that Tucson Water can find a Permit Tracking System that has already been implemented by a municipality, but it is unlikely that any system located will perfectly meet the needs of Tucson Water.

Tucson Water could embark on the development of a customized Permit Tracking System, but this is likely to be expensive and will cover work already done many times by other municipalities, utilities, agencies and commercial vendors.

If a Records Management System is implemented, the actual permits and communications could be tracked in the system. However, the Records Management System would lack the capability of tracking the permitting process that would be included in a Permit Tracking System.

Recommendations - Tucson Water should clearly define their exact requirements for a Permit Tracking System and should then explore the availability of commercially available software that can be installed and customized either by the vendor or a consultant to meet the specific needs of Tucson Water. The importance of clearly defining the requirements before exploring options is especially important in this situation since there is such a vast set of options for meeting this requirement.

Benefits - A Permit Tracking System would facilitate the submission, tracking and response to communications regarding permits. It would also track the start and end dates of the permits. A system of this type would allow appropriate personnel to quickly review the status of existing and requested permits.

Costs - Since Permit Tracking Systems are already available commercially and there is a limited number of likely users for this functionality within Tucson Water, it is likely that an inexpensive solution can be identified. To control the costs of meeting this requirement, Tucson Water could document the requirements internally and then include the requirements in an RFP. The cost of purchasing and implementing a system of this kind will depend on the detailed requirements specified and how much of the implementation can be done in house. Clearly, limiting the customization of the application chosen will reduce the cost and complexity of the implementation.

- **Human Resources Information System**

Description – The Human Resources group manages critical data for the Department and they are currently using a number of different systems to track this information. These systems include the City of Tucson Human Resources Database (AHRS), which includes personnel records, and Training Wizard 97. Other critical data is maintained in a Microsoft Access database developed specifically for the Business Services section. Currently, these systems are not integrated and it is difficult to keep them all up to date and in agreement. The Business Services section appears to be lacking resources both in personnel and in hardware to make these systems work as needed. The two divisions accessing this database have been Operations and Planning & Engineering.

The training data is maintained by third party software called Training Wizard 97. This software is SQL-based and accesses the personnel database. Training Wizard tracks what training an individual has completed. Safety training, license and certification requirements are maintained in the database. Training representatives from each division could share the Training Wizard, but currently only the Training Coordinator and one secretary use it due to installation barriers. Operations utilized the software continuously until the Operations representative that knew how to use the software was transferred. Training information input from Operations has not been updated since and is backlogged. The Division Administrators and the Business Services Division identified these requirements.

Options - This requirement is for optimizing and integrating existing systems. This work could be done either internally by Tucson Water personnel or by contract with outside vendors.

Recommendations - This integration is best accomplished by internal staff since it involves existing systems that are running. A consultant will have a significant learning curve to understand the existing systems and organizational barriers to information exchange. In order for an internal implementation to work, the Tucson Water database administrator will need to map all the databases involved and ensure that they are compatible. Batch processing of existing data will also need to be evaluated to see how easily data can be transferred between systems. Data flows should be mapped and potential users should be identified. A plan should be developed for who and where updates will be performed and how the other databases will be updated. The ideal situation would be for pertinent information to be located in a single database with clear responsibility for data entry, quality assurance and reporting. Since these data are sensitive, security considerations are critical. Access to these databases must be vigorously restricted.

It is likely that integrating these information systems will require additional staff. The required skills for performing this work is currently available within the IS section but they are understaffed to focus on these types of issues.

- **Material, Data, Handling Sheets**

Description - These are sheets that specify the characteristics of materials used in the utility that may be dangerous or require special handling. These sheets are required to be posted and available to personnel. Historically, these have been maintained by the Safety officer and have been distributed to appropriate locations in hard copy. They have not been updated in some time and there is no official tracking process for the forms, update or distribution.

Options - If the Records Management System is implemented, the handling of these data sheets can be optimized. Records management could be used to track these documents, including version control and access to online documents. Links to databases about each compound could also be built into each document allowing personnel to quickly locate associated material available either on the Local Area Network or on the Internet.

If the hard copy forms are required, they could be printed and a separate database or spreadsheet would need to be kept of the last time the forms were updated. The versions could be easily tracked in the Records Management System.

If the Records Management Software is not implemented, then a small database should be developed that can be web-enabled. The database should list the materials and the files available with the data sheets. These can be made available in a web interface to appropriate personnel within the utility. Any version tracking done using this strategy will have to be done manually. Hot links to actual information sheets can be easily implemented. This type of an implementation would fit in very well with the Intranet initiative described previously.

Recommendations - If a Records Management System is implemented then this functionality should be tracked within that system. But if the implementation of Records Management System is not planned in the near future, then the small database necessary to meet this need should be developed in Microsoft Access or in Oracle. The information in the database should be made available online to allow employees to locate material handling and data sheets that are stored online.

Benefits - These information sheets are important to protect Tucson Water employees, water quality and emergency response personnel that might be called in during a spill or an emergency. Development of a system and online archive of data sheets will protect Tucson Water from possible regulation violations and legal action that could result if current sheets are not available.

Costs - If a Records Management System is implemented then applying this functionality within the records management software will require only the time of Tucson Water personnel.

If a small application needs to be developed it can be developed inexpensively by existing personnel, contract or by hiring a summer intern.

EAST BAY MUNICIPAL UTILITY DISTRICT

UTILITY MANAGEMENT

The East Bay Municipal Utility District (EBMUD) supplies water to parts of Alameda and Contra Costa counties on the eastern side of San Francisco Bay in northern California. Approximately 1.3 million people are served by EBMUD's water system in a 325-square-mile area extending from Crockett on the north, southward to San Lorenzo (encompassing the major cities of Oakland and Berkeley), eastward from San Francisco Bay to Walnut Creek, and south through the San Ramon Valley. EBMUD is a publicly owned utility formed in 1923 to provide water service and expanded in 1951 to include wastewater treatment. EBMUD has a seven-member Board of Directors publicly elected from each of seven wards (see Figure C.4) within the EBMUD service area.

EBMUD has approximately 2,000 full-time employees under the administrative direction of the General Manager and management staff. The primary organizational departments of EBMUD are:

d. Office of the General Manager
e. Water and Natural Resources
f. Operations and Maintenance (Drinking Water Systems)
g. Engineering and Construction
h. Administration
i. Finance
j. Wastewater

To focus on water quality related information needs, most groups involved in this assessment were part of the Operations and Maintenance Department. The primary exceptions were the Information Systems Department, which is part of Finance, and Water Supply Improvements, which is part of Water and Natural Resources.

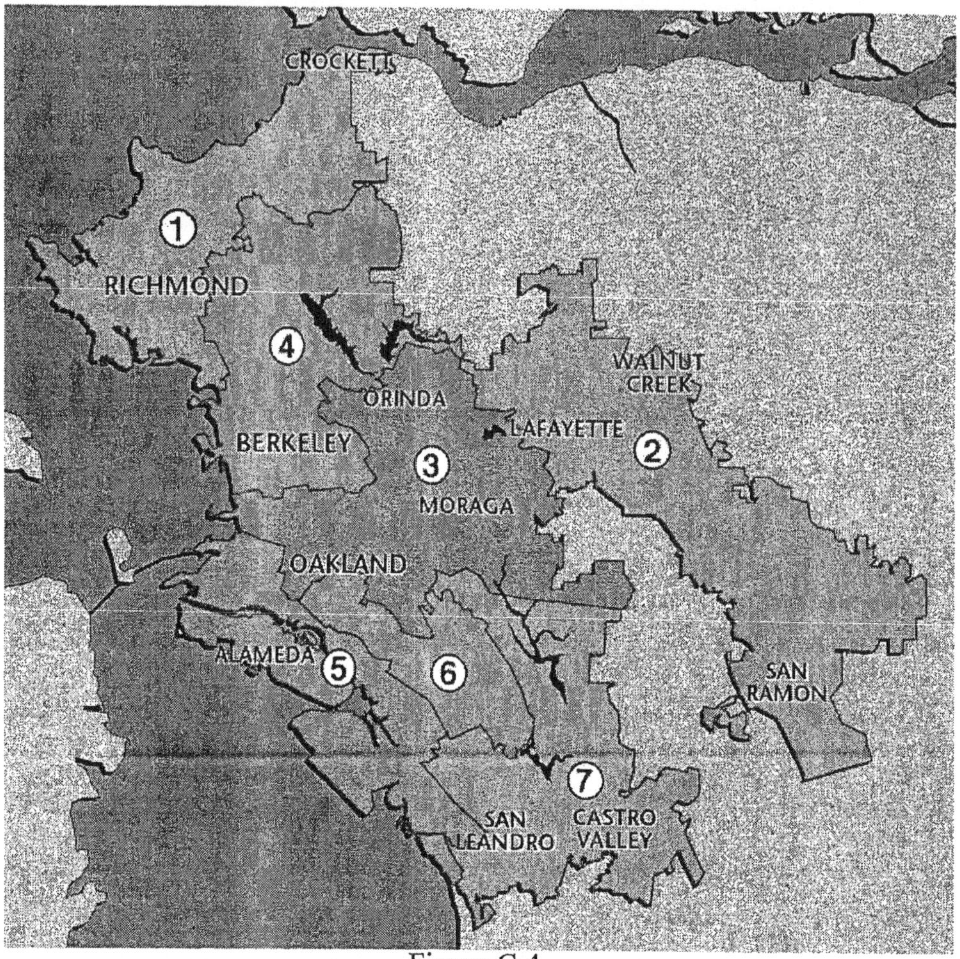

Figure C.4

Seven Wards comprise EBMUD

WATER TREATMENT PLANT OPERATIONS

EBMUD's six water treatment plants are capable of filtering and processing more than 400 million gallons of water daily (mgd). Currently, the average daily demand is approximately 215 mgd. The water treatment plants are located in Upper San Leandro, San Pablo, Sobrante, Orinda, Lafayette and Walnut Creek. The Orinda Water Treatment Plant has the largest output, with a capacity of 175 mgd. This plant serves all or parts of Alameda, Albany, Berkeley, El Cerrito, Emeryville, Moraga, Oakland, Orinda, Piedmont, Richmond and San Leandro. The other plants supply water in varying amounts to the balance of the district.

PRELIMINARY DATA COLLECTION AND INTERVIEWS

There were 19 questionnaires completed and returned to the project team prior to the site visit. The interview and reporting process for EBMUD was scheduled over a four-day period – March 26th through 29th. Interviews were interspersed over the first two and a half days. The project team worked the afternoon of the third day and presented the findings the morning of the fourth day, Thursday, March 29th. The goal of this reporting out session was to validate the information collected, facilitate further discussion and present preliminary results of the information collected. Table C.14 lists the participants who completed questionnaires and participated in the focus group meetings.

INFORMATION TECHNOLOGY

EBMUD's Information Systems Department (ISD) consists of approximately 75 people to provide support to the 2,000 employees. ISD is further subdivided into three groups:

- **Database Administration** – Supports the District's database and geographic information assets.
- **Data Center** – Provides hardware, network, and systems software services and support.
- **Applications Development** - Maintains the customer information, billing, financial, and payroll packages.

Many EBMUD information systems have been developed internally, especially database and GIS applications. The standard desktop application is Microsoft Office 2000. While Microsoft Access is available for database management it is not widely used for data storage. Most database applications are Oracle-based systems using Cold Fusion to produce web-enabled access to the information. The Utility's intranet is an integral component of data access for most groups. ISD has a mission to make data available and accessible to all employees.

Table C.14 Stakeholders

Stakeholders Name	Department	Questionnaire	Interview
Allen Goodson	Asset Management	X	X
Hilary Anderson	Asset Management	X	X
Gus Duarte	Asset Management		X
Lee Vestal	Distribution		X
Alvin Tong	Engineering	X	X
William Kirkpatrick	Engineering	X	X
Xavier Irias	Engineering		X
Hubert Lai	Engineering Services		X
Ali Sheikholeslami	Engineering Services	X	X
Andrzej Wilczak	Engineering Services	X	X
Andy Datlen	ISD	X	X
Colleen Swain	ISD	X	X
Herb Seidell	ISD		X
Joan Smith	ISD	X	X
Mike Vandaveer	ISD		X
Robert Frey	ISD	X	X
Bill Elgas	Lab Services		X
Julia Hasne	Lab Services		X
Laura Lazzell	Lab Services	X	X
Jeff Murray	Maintenance Support		X
Leonard Pristia	Maintenance Support	X	X
Tim Collins	Maintenance Support	X	X
William Kerr	Maintenance Support	X	X
Bryan Dilts	OP/NET	X	X
Diosdado Hernandes	OP/NET		X
Kevin Wong	OP/NET	X	X
Steve Kachur	Operations	X	X
Richard Sykes	O & M Management		X
Ron Hunsinger	O & M Management	X	X
Brian Marler	Operations Engineering	X	X
Joe Young	Operations Engineering	X	
Don Jackson	Pipeline Construction		X
Alexander Coate	Regulatory Compliance	X	X
John Schroder	Regulatory Compliance		X
Robert Berger	Water Quality Office	X	X
Andi Enos	Water Supply Division		X
Cliff Threlkeld	Water Supply Division	X	X
Jim Pierner	Water Supply Division		X
Tom Heaney	Water Supply Division		X
Jeanette Madden	Water Supply Engineering	X	
Hasan Abdullah	Water Supply Improvement	X	X
Mark Bluestein	Water Supply Improvement		X
Mark Williamson	Water Supply Improvement		X
Mike Tognolini	Water Supply Improvement		X
Jim Smith	Water Treatment Section	X	X
Mary Hicks	Water Treatment Section	X	
Rod Jung	Water Treatment Section		X

The Laboratory uses a commercial LIMS produced by LabVantage that meets the needs of the laboratory. There is a full time technician assigned to support the LIMS. In addition to the central lab, there are process control labs at the plants. The process control personnel enter water flow into the LIMS, but most of the process control data is entered into a separate System Water Quality database.

Other systems maintained by ISD include, PeopleSoft applications for Human Resources and Financial systems, HEAT which is a Help Desk Application, Work Order Systems, and Maximo.

The Systems Programming Group handles the backups. They have implemented daily and weekly backup routines for enterprise databases and file servers. Disaster plans include the storage of backups offsite.

SCADA SYSTEM

EBMUD currently has a SCADA system in place for control and data acquisition. The Human Machine Interface (HMI) at each water treatment plant is manufactured by Fisher-Aimax and the operating system is PC based. The SCADA system is connected via a Wide Area Network (WAN) and is referred to as the OpNet system. The HMI system (OpNet) has approximately 16,000 tags (inputs and outputs) with five area control centers and a central control center. The information on the OpNet system is real time data and is stored for a period of two weeks. The Water Treatment Operations personnel have access to the OpNet system and can monitor all points from each water treatment plant. The OpNet system is connected to a Unix based HP-UX system (OSCI) for long-term storage of compressed data. The OSCI data is compressed and is stored for a period of at least three years. The OSCI data can be accessed by district personnel and is the interface for most users that are not in Water Treatment Operations. The water treatment plant controls are primarily Allen Bradley Programmable Logic Controller (PLC) using DF1 and Data Highway Plus communication protocols. The RTUs (Remote Terminal Unit) communicate via a proprietary HP protocol. The OpNet group is in the process of implementing a new SCADA system. The HMI OpNet system will be replaced with a Metso Automation system (OASys). The OSCI system will be replaced with an Aspen InfoPlus.21 system. The new systems will operate on a Windows NT platform. The OpNet group is also performing a pilot project with IndX XHQ software that utilizes "hooks" into multiple disparate databases on a WEB based portal platform.

OBSERVATIONS

EBMUD has made great strides in the past few years in organizing an effective and comprehensive program for the capture of most of the raw data that is essential to the organization. Following the capture of the raw data however, the quality assurance and control of results is not consistent. Many of the paths for data capture include comprehensive and effective quality assurance and control while other processes do not. Similarly, some of the aggregation of results is done effectively while other aggregation processes are *ad hoc* and not performed consistently. This means that the same data capture process might be completed using many different approaches for the same data.

Raw data capture within EBMUD ranges from adequate through excellent. A significant portion of the data collected is not shared or generally accessible. While there are limitations in the use of on-line data, EBMUD leads the participating utilities in the study with the number and extent of data that are accessed on-line. The Asset Management Group within the Operations and Maintenance Department has been assigned the responsibility for organizing and supporting the dissemination of data both within and across sections of EBMUD. This assigned role is temporarily on hold while other priorities are addressed. However, it does appear that EBMUD is heading in the right direction. A separate group such as Asset Management committed to a water quality facilitation role rather than a role of the ownership or stewardship of the data is a positive step.

EBMUD has made unique strides in making SCADA information available to many users within the utility. The OpNet data is available to a wide range of users and OSCI is used as a data historian and disseminator for regularly aggregated and quality assured data. This has had significantly positive results in the availability of SCADA data to a variety of users within the utility. The planned upgrade to the Metso Automation's OASys Distributed System (NT) and Aspen Tech's InfoPlus 2.1 (NT) appears to be another step in the right direction for further accessibility to SCADA data. This upgrade should address some of the concerns expressed by users of the SCADA data related to limitations of the current data that are captured by the historian. Specifically, the lack of detailed information and some of the difficulties in downloading SCADA aggregated data to other applications for further analysis and use should be addressed.

INFORMATION MANAGEMENT OBSERVATIONS

The general impression of the majority of the users from EBMUD is that the Information Systems Department (ISD) is doing an outstanding job supporting and continuing the development of information capability within EBMUD. While there are some that disagree with the majority, they still acknowledge that ISD is doing an acceptable job. This perception again ranges from adequate through excellent. Those dissenting from the excellent view seem to share the opinion that the lack of an outstanding grade is due to the fact that ISD is understaffed and is a victim of its own success. ISD has made great strides in the last few years in improving both infrastructure and processing of information. However, they have not been able to apply their resources to all groups within the utility and their success has raised expectations throughout the utility. ISD has succeeded not only in improving the processing of information but also in the deployment of aggregate information and directories to an excellent web site.

Observations about Information Management at EBMUD include:

- There is major disagreement within the utility regarding how much is left to be accomplished by ISD in order for EBMUD to have a comprehensive information system. Many of the ISD personnel feel that little needs to be done; others within the utility feel that a great deal needs to be done.
- One issue identified consistently is the lack of communications regarding the available systems and functionality of those systems. Currently a summary of

information resources is not available either on-line or off-line (i.e. there is no directory of available information or systems).

- There is also a general perception that often when new systems are developed, the stakeholders included in specifying the requirements are far too limited. We observed a wide range of approaches to performing requirements analyses. Some of these were comprehensive; others were significantly limited. Perception of limited input may also be caused by the fact that often the requirements analyses are performed by stakeholders with a vested interest in the outcome of the requirements analyses. We will discuss approaches for alleviating this perception in the Recommendations section.
- Another major deficiency that we observed is the lack of a comprehensive master plan for information systems development over the next few years. Such a plan would present a framework within which individual systems could be developed promoting the ability to integrate systems into a single comprehensive information system. Currently the direction of the information systems within EBMUD is driven by grass root efforts. As compelling needs are identified, they are addressed on a section-by-section basis. We were not able to identify a single high-level manager within EBMUD who is acting as the champion for integrated information system development. The possible exception to this would be the Manager of Water Systems. This could have a significant effect on the management commitment and the integrated nature of decisions made regarding the information system development path.

DATA QUALITY

Many concerns were expressed during the requirements analyses concerning the data quality. Depending on the type of data being collected, there is a perception that some of the data are adequately quality assured while others are not. The project team's perception of data processing within EBMUD is summarized in Figure C.5. In the data hierarchy described in Figure C.5, it seems that EBMUD is doing extremely well at the lowest level of data capture and access. Some of the quality assurance and control that should be performed is getting done well while some of it is not being done adequately. Therefore we show the first level in green while part of the second level is green and part of it is yellow. The remaining levels of data aggregation, interpretation and development of knowledge is being done well in some areas but not well in others therefore the remaining cells are yellow.

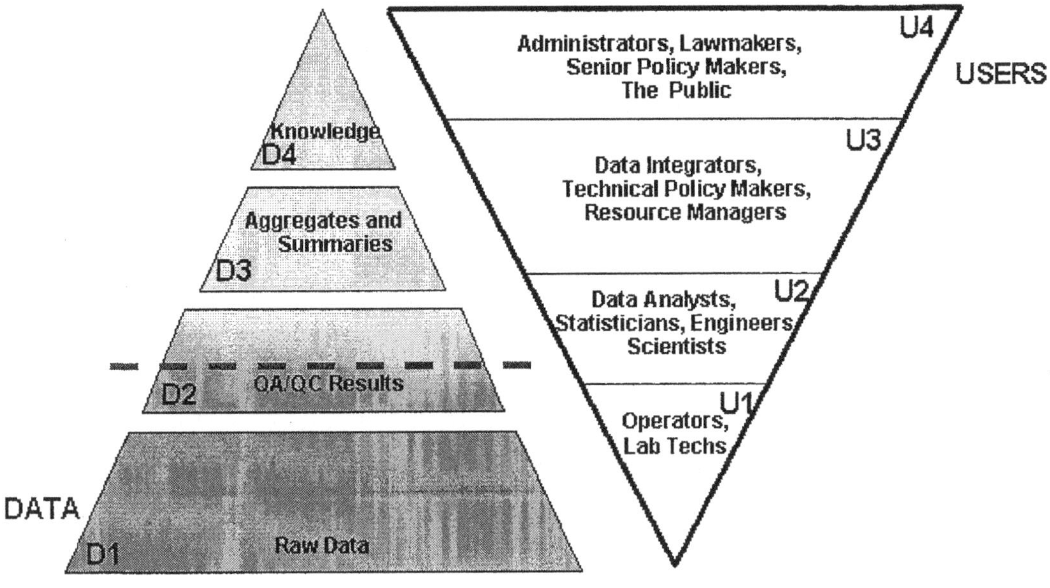

Figure C.5

Knowledge management at East Bat MUD

- One valid, major concern expressed repeatedly is that when data are being extracted and integrated from a wide range of raw data, the results can be compromised due to disparate levels of quality control. Much of the integration that is currently done requires extensive filtering of data and manual inspection of data to ensure that data elements can logically be combined together. This also highlights a significant issue regarding the documentation of data collection methodologies and the context of the data that are available on-line. There is not much documentation about the data or about methodologies available on-line. These issues result in increasing the complexity of the available data and the integration of the data (shown as level D3 in Figure C.5) because of the lack of unambiguous data and adequate documentation to relate different sample types. In some cases, concerns were raised regarding the inappropriate use of fields within databases. If standards were implemented and required, some of these problems could be alleviated. An example of this is in the

LIMS field for sample location where the code MISC (miscellaneous) has been entered into the field in the database. Detailed location information has then been entered in the comment field. This results in an inability to query the database for specific locations and capture all of the available information. It requires any user to then search through the comment fields to determine if a particular record that has been labeled as miscellaneous is, in fact, a member of a sub-set that is required for the data analysis. Further, if the data are appropriate for inclusion in the analysis then the locational information need to be entered manually and integrated in with the data record.

- Similarly, data critical to the use and interpretation of SCADA data from on-line instruments is currently not maintained in an electronic system. For example, as calibrations are performed, the data are hand entered onto paper documents, which are then stored in file folders. This absence of critical metadata creates an inability to search and quality control the electronic results used during any query without going back to the original paper records related to the calibrations.

- The participants in the requirements analysis presented inconsistent perspectives on the status of standard naming conventions and normalization of data fields. This means that not all people within the utility are aware of standard naming conventions for locations, materials and analytes. In addition, there was some disagreement regarding whether all people entering data for a particular field are entering the same quantities in known units measured in the same way. This lack of agreement on standard naming conventions is of great concern. It indicates that there is, in fact, more than one set of standards that have been generated within the utility. There are users who do not have confidence that the labels, location names and analyte names are being consistently used throughout the utility. If these standards are not universally used, then sub-sets of any on-line databases may result in inconsistent data being included in data aggregates which are then analyzed, summarized and included in reports used for decision makers, regulators and for compliance recording.

- A general requirement for improved records management emerged from many of the focus groups. The needs expressed include keeping track of both electronic and paper records in an orderly fashion. However, a compelling requirement for records management was not identified. Each individual or section felt they could perform his or her work without records management. However, records management was consistently identified as an important consideration for the future. The Manager of Engineering Services is a champion within the user community who is willing to take responsibility for investigating and introducing records management into EBMUD.
- The focus groups identified an inconsistency in the processing of similar data sets. Two examples were the Hydro Lab data being processed differently by different groups within EBMUD and the handling of treatment operational data from different facilities within EBMUD.
- There is little data tracking for the vertical Hydro Lab profiles. Hydro Lab records are tracked by file name and are not standardized between different groups.
- Some modeling activities have been identified within EBMUD. In general these models are run as one-time outputs and are not stored. The hydraulic model used by EBMUD employs OpNet inputs (SCADA raw data). These inputs are batched and loaded into the model. There is no direct database interface to the model. Other models are used to study processes in the reservoirs; these are dynamic models. Data are either input from measurements made or to evaluate different scenarios. Other modeling efforts were attempted using data from MAXIMO to perform cost efficiency analyses. It was found to be very difficult to download the data out of MAXIMO and load it into the models and spreadsheets for evaluation.
- Users of the on-line databases indicated there is a lack of robust *ad hoc* query capability. With the difficulties identified above in sub-setting data, the inability to perform robust *ad hoc* queries presents yet another barrier for the integration of information by a wide-range of users within EBMUD. This results in users downloading from the database into other analytical tools (such as spreadsheets) and manipulating the data based on individual assumptions that may or may not be

consistently held throughout the utility. The potential result of these disparate assumptions could be that two analysts addressing the same question might sub-set the data differently, aggregate the data differently and come up with two distinct and different answers for the same question using the same data.

- Currently the knowledge of the data quality and both the limitations and appropriate uses of the data reside with the data owners or those who use it on a regular basis. The actual authoritative usefulness and limitations of the data is difficult to access without inputs from a number of parties within EBMUD. Again, this results in an inability to perform analysis on the data by a wide range of users within EBMUD. It also results in an inability to use the data in any way other than those originally envisioned.

DATA PROCESSING OBSERVATIONS

As indicated earlier, EBMUD has made great progress in the capturing, processing and use of on-line data. In areas where ISD have focused, there are robust systems for capturing, quality assurance and processing data. There are currently standard operating procedures for integrating new and updating existing spatial data for engineering functions. The availability of up-to-date spatial data is invaluable and is fairly unique within the Drinking Water Industry. EBMUD also has made a huge investment in the development of current information on system infrastructure. The available data are near to up-to-date and again this is unique within the Drinking Water Industry. The availability of these data is impressive and very valuable in operations, management and future planning for EBMUD.

Some observations that arose in discussions relative to data processing include:

- In some cases where systems have been customized and developed by ISD, the systems have fallen into disrepair and have not been updated regularly because of ISD moving on to other projects. When ISD completes a particular phase of the project and turns the system over to the intended users, the systems have worked adequately, however, lack of a maintenance and management plan for the systems has resulted in the systems not being kept up-to-date and maintained to meet the ever changing needs of the sections and divisions. In some cases, the systems have therefore been abandoned. The opinion of the users is that this has occurred because of lack of resources within ISD to both maintain old systems and continue development of new systems.

- The SCADA system is not currently programmed for the data cleanup and capture requirements under the interim enhanced surface water treatment rule (IESWTR).
- The processing of data from its raw form into aggregates and the development of reports is performed through a combination of automated and manual processes. The EBMUD environment is heavily "Excel-centric" for data reporting and data analysis. A great deal of the integration that is currently done with existing data within EBMUD is "People-centric". This results in simple failures when these key people are out of work for vacation or illness. In many cases, there are multiple backups for these key people minimizing the concern that critical tasks cannot be performed when key people are not accessible. This concern though minor could be further mitigated through the automation of reporting and automatic aggregation of results. If standard reports and summary aggregates were automated then more people would be able to generate required reports with less of a reliance on a few key individuals.
- The majority of the data collection and processing is oriented from a Divisional perspective rather than a utility perspective resulting in insufficient documentation regarding the methods of sample collection and the limitations of the analyses and output results. It also results in information systems that are vertical within a particular division or section (silos of information and systems). These data silos are often unique to a particular program, section or division making the sharing and use of the data by others very difficult.
- The desire to upgrade the hydraulic model in order to help it predict water quality has been articulated. There currently is no strategy or plan for doing this. Discussions regarding the interfacing of the model with the databases both on the

front-end for parameterization and input as well as on the back-end, so that once models are run the results can be stored in databases was discussed.

FIELD DATA OBSERVATION

A strong desire was expressed during many of the focus groups' meetings for field personnel to have access to computer systems or interfaces for recording, entering, and accessing data. Many of the field staff indicated difficulties in accessing the information while they were in the field resulting in a return trip to EBMUD facilities to access and download the necessary data.

Any field systems developed for EBMUD field personnel need to be implemented in the proper scale and also need to be user-friendly. The scale issue relates to the type of computer systems deployed for field staff. In some cases, field personnel need hand held (e.g., palm pilots) or lap top computers in order to make them useful in the field. Complex large systems in the field may not alleviate any problems. In all cases, field systems should include single entry strategies wherever possible, bar codes and infrared mobile tags (e.g. speed pass) and direct input or uploading of data from instruments in the field. Currently, a significant portion of the field data collection is self-contained within a particular division or section's system. Often times these data are recorded on paper and stored in file cabinets. Field collected data are very seldom integrated into the overall system. The exception to this would include some of the GIS data for which standard operating procedures have been developed.

GENERAL OBSERVATIONS

During the focus group meetings, a number of overarching observations emerged that are presented below.

- There is a perception within EBMUD that program or project related data are more safely managed when they are isolated from the integrated system. It is assumed that once data are stored in a central system, the data are somehow at risk or of lower quality than when managed in the originating group. This results in different standards and difficulty in integrating databases due to the unique implementation of each project within a division or a section.

- Whereas EBMUD has made significant progress in the storage of some of the raw data, it is still difficult to access data due to limited data querying capabilities as well as widely varying QA/QC. This results in difficult and often incorrect integration and aggregation of data. Often, critical data which are available on-line cannot be incorporated into integrated assessments in a simple manner due to documentation discrepancies and difficulty in accessing and using the data. A lack of detailed understanding regarding the appropriate uses of the data further complicates access and integration of information into common systems.

- Once data are available in common systems, EBMUD does not have many graphical tools or robust analytical and visualization tools for viewing data. Some of these tools are used and maintained by individual divisions or sections; however, they are not widely available to many users within EBMUD.

- EBMUD does not have broad support for applications beyond the standard Microsoft Office suite of software tools. This could be part of the reason that EBMUD is so "Excel-centric" for data extraction analysis and reporting.

- Exchange of data to and from external sources appears to occur on an as needed basis. These exchanges do occur and in many cases have been established and are done efficiently and in other cases the exchanges are ad hoc and are not done efficiently. Specific examples of exchange of data include water resources and climatological data. These exchanges occur both from EBMUD to the outside as well as from the outside into EBMUD.

- The GIS capabilities within EBMUD are under utilized and are perceived by the majority of users as a difficult application. This suggests that while some interfaces to the general user community do exist, in general, easily accessible non-GIS specialist interfaces have not been developed or have not been communicated to the user community. An exception to this observation is in the Engineering Section where a dedicated information systems development GIS staff member resides and is committed to the Engineering Section.

BARRIERS

In this section we will identify and summarize barriers that were identified. Many of these barriers have been discussed in the context of observations above.

- The absence of a comprehensive integrated central information system and standards for that system is a major barrier to the user information within EBMUD. The fact that the data are of unknown quality and are often not adequately documented further exasperates the use of the data. These problems result in extreme difficulties in accessing, sub-setting and aggregating data that has been collected in various programs. Therefore sub-setting the data, generating integrated reports and arriving at integrating summaries is very difficult. The lack of standards for naming conventions both in parameter names as well as in analyte codes and location codes, further complicates the integration of information.
- The lack of a comprehensive directory of information that informs users of the availability is a serious problem.
- As discussed earlier in observations, the perception that GIS is a difficult application, which is hard to learn and to use, leads to its underuse.
- Communication between different functional groups has not been effective regarding data about availability and opportunities for collaboration. When information systems are developed within a division or a section, the requirements of other groups within the utility are often not taken into account resulting in systems that may have one specific need but do not meet a wider need within the overall Drinking Water Utility.
- Conversely, the lack of clear priorities in the development of information systems within the divisions and sections has led to a lack of confidence by ISD in the ability of some groups to clearly articulate their information system and data content

requirements. This results in systems sometimes not being developed according to the users specification because of an incorrect perception that the requirements are not clearly defined. This has also led to a lack of broad stakeholder perspective in the design of the information systems because the appropriate variety of stakeholders are sometimes not included in the requirements analysis due to a lack of confidence that the stakeholders will be able to converge on priorities, content and functionality requirements.

- The designation of Asset Management as the advocate for data management within a number of groups has been positive but the lack of advocacy for some groups within the water quality arena, specifically processing engineering, distribution systems and water supply, has led to a lack of expertise and utility of data within these groups.
- Many standard reports that should be automated have not been, resulting in some reports being put together manually.
- There is a deficiency within the standard suite of tools within EBMUD for data analysis and graphical presentation of data. This is a barrier to the use and communications of results within the utility.
- There has been inconsistent system rollout and training support to achieve optimal system utilization. Sometimes the rollout is limited to the targeted functional group where additional broad-based training might increase the use of particular systems. In addition, ISD has not been able to continue robust support of applications once they are completed with training and system rollout. It would be appropriate for the systems to be supported more adequately as well as to regularly hold training sessions. It is our understanding that this is a plan with Asset Management taking an active role in data advocacy.

RECOMMENDATIONS

This section includes specific recommendations that the project team thinks would be helpful for EBMUD to improve the capture, use and summation of information.

- The planned information sub-committee needs to be initiated as soon as possible. This sub-committee can act as the management champion for information systems. The information sub-committee should act as the advocate for information management in general. This sub-committee should prioritize implementation of information system needs and should act as an advocate for the required resources to meet these needs. Part of the sub-committees task should be the initiation of standards development. A requirements analysis should be conducted to define which standards need to be focused upon and these should be addressed as quickly as possible. Some clear candidates for immediate standards include location conventions, coding for quality assurance and control, attribute naming conventions for database development and the development of documentation standards for data included in the EBMUD integrated information system.

- EBMUD management should make clear policy statements regarding the roles and responsibilities of the Asset Management Group. We heard a number of perspectives on this ranging from supporting the development of information systems through acting as an information bureau capable of supporting different users and accessing data and in grand integration of data across many data sets within the utility. Until Asset Management's role is formally defined, there will be varying levels of acceptance of their role in water quality information management.

- The division or section managers should formally extend the role of information system advocacy to those users within their groups that have accepted these responsibilities in an *ad hoc* manner. Those who have become the de facto point of contact for supporting data access and integration within the group should be formally recognized in these roles and should begin developing formal ties to those within Asset Management who are given that responsibility.

- Within each division and section, managers should begin meeting with staff to understand overall information requirements and to begin prioritizing these information needs from an organizational perspective.

- A requirements analysis should be conducted overall for the utility to determine what new systems are needed. This effort should include representatives of all stakeholders and should be facilitated by an outside moderator to ensure that all stakeholders are integrated into the process. No stakeholder should be facilitating these meetings to prevent biased priorities for required systems. The disinterested facilitator may come from within EBMUD. In situations where the requirements become complex, a third-party facilitator should be considered.

- The Asset Management Group should begin developing directories of the available data sets. These should be loaded into an EBMUD intranet database and should be regularly updated by the Asset Management Group. A data dictionary should also be developed which lists the names of the data sets (also called entities) and the column headings (also called attributes) for each data set. Each data set should have information regarding the content and the on-line location if the database is available.
- Current databases do not include the ability to track information at different depths. For example, the reservoir data can only be stored for a specific location on the surface. This needs to be reconsidered and depth should be incorporated into the sample location identification scheme.
- EBMUD should begin exploring the use of the latest generation of GIS applications, including Internet mapping services (ArcIMS). GIS should become a standard front-end for display of environmental data including water quality data. The in-house GIS capabilities are currently underutilized.
- EBMUD should evaluate the expansion of records management capabilities to technical groups within EBMUD. The Engineering group currently uses the Records Clerk to request drawings that are being maintained and managed by EBMUD. When the Clerk scans the Mylar drawing to print it, the electronic scan is then deleted rather than possibly being saved into an electronic document management system for reprinting without an additional scan.
- ISD and Asset Management should initiate dialogue with those groups for which systems have been implemented but are currently unused; specifically, complaint tracking and field sampling.

- The groups doing hydro-lab data casts, capture and analysis should meet with members of ISD and Asset Management to evaluate the possibility of an integrated solution to hydro-lab data processing storage and dissemination. Currently the data from the hydro-labs are dealt with in two completely independent manners.
- The Asset Management Group should be continuously trained to ensure that they have a current perspective and current knowledge of the available tools and information.
- EBMUD is ready for the implementation of Knowledge Management. Knowledge Management systems could be used to facilitate some of the quality assurance and control of data; it could be used to automate data filtering and a generation of standard reports; it also could be used to automate presentation of data aggregates using web-enabled technologies. Knowledge Management could also be used to organize data on the Internet for different groups of users within EBMUD.
- Currently, ISD provides a web-based E-Query tool for general queries of data, but it only operates on a single table or view that has to be predefined by ISD. EBMUD should acquire an ad hoc query tool for existing information systems. This tool should allow users to generate queries using screen-oriented entry specifying the fields that the data should be subset on and the range of values, which are desirable. Tools purchased should include the capability of saving a query to be resubmitted at a later date.
- The Water Treatment Report Manager (WTRM) system should be reconsidered for modification and expansion. The program should be rebuilt to District standards and the expansion should focus on expanding the user base, especially for the process-engineering group.
- An on-line instrument maintenance and calibration log should be included in the modified WTRM system. All training should include direction for the use of this log, both from a data entry and a data use perspective. This will result in better quality assurance of data captured from on-line sensors. The WTRM system already has the capability of employing data ranges. These should be populated, maintained

and turned on. In addition, other QA/QC automated checks should be created for data entry and analysis of the WTRM data.

MASSACHUSETTS WATER RESOURCES AUTHORITY

UTILITY MANAGEMENT

The Massachusetts Water Resources Authority (MWRA) is an independent public authority that provides wholesale water and sewer services to its customer communities, and funds its operations primarily through user assessments and charges. MWRA was created by the legislature in 1984 and inherited operations and facilities in 1985 from the Metropolitan District Commission. MWRA is governed by an 11-member Board of Directors who are appointed by the Governor or directly or indirectly by elected officials in MWRA customer communities.

MWRA's 1,570 employees serve approximately 2.5 million people and 5,500 industrial clients in 61 customer communities collectively between water and sewerage services. Table C.15 lists the current communities served, at least in part, by MWRA.

Table C.15
Communities Served by MWRA

Water Only	Sewer Only	Water and Sewer	
Chicopee	Ashland	Arlington	Newton
Leominster***	Braintree	Bedford*	Norwood
Lynn**	Burlington	Belmont	Quincy
Lynnfield WD	Dedham	Boston	Revere
Marblehead	Hingham	Brookline	Somerville
Marlborough*	Holbrook	Cambridge*	Stoneham
Nahant	Lancaster	Canton*	Wakefield
Northborough*	Natick	Chelsea	Waltham
Peabody*	Reading	Clinton	Watertown
Saugus	Randolph	Everett	Wellesley*
Southborough	Stoughton	Framingham	Winchester*
S. Hadley FD 1	Walpole	Lexington	Winthrop
Swampscott	Westwood	Malden	Woburn*
Weston	Weymouth	Medford	
Wilbraham	Wilmington	Melrose	
Worcester***		Milton	
		Needham*	

*indicates partially supplied water by MWRA

**water supplied to GE only

***MWRA is emergency back-up water supply

WATER TREATMENT PLANT OPERATIONS

In the year 2000, demand on the MWRA system was 252 million gallons per day (mgd), including a temporary 14 mgd demand from the City of Cambridge. Based on a detailed analysis of major factors affecting water demand, MWRA projects that system demand will remain in the 240-260 mgd range. MWRA estimates it has an active capacity of 478 billion gallons and an emergency back up capacity of 11.2 billion gallons. Table C.16 shows the individual reservoir storage estimates that make up MWRA's supply.

Table C.16

MWRA Capacity

ACTIVE CAPACITY: 478 billion gallons	
Reservoirs 477 billion gallons	Quabbin Reservoir: 412 billion gallons
	Wachusett Reservoir: 65 billion gallons
	Norumbega Reservoir: 205 million gallons
Covered Storage Facilities 65 million gallons	Nash Hill Covered Storage Facility: 25 mg
	Fells Covered Storage Facility: 20 mg
	Loring Road Covered Storage Facility: 20 mg
EMERGENCY BACK-UP CAPACITY: 11.2 billion gallons	
Source Reservoirs 8.4 billion gallons	Framingham Res. #3: 1.2 billion gallons
	Sudbury Reservoir: 7.2 billion gallons
Distribution Reservoirs 2.7 billion gallons	Spot Pond Reservoir: 1.9 billion gallons
	Chestnut Hill Reservoir: 522 mg
	Fells Reservoir: 64 mg
	Weston Reservoir: 200 mg

PRELIMINARY DATA COLLECTION AND INTERVIEWS

There were 13 questionnaires completed and returned to the project team prior to the site visit. The interview and reporting process for MWRA was scheduled over a three-day period, December 12th through 14th. Interviews were interspersed over the first day. The project team worked the second day and presented the findings the morning of the third day, Thursday, December 14th. The goal of this reporting out session was to validate the information collected, facilitate further discussion and present preliminary results of the information collected. Table C.17 lists the participants who completed questionnaires and participated in the focus group meetings.

Table C.17
Stakeholders

Stakeholders Name	Department	Questionnaire	Interview	Presentation
Betsy Reilley	Lab Staff/Quality Assurance	X	X	
Winsor Sung	Lab Staff/Quality Assurance	X		
Peter Ralston	Data Management Staff	X	X	X
Clare Kirk	Data Management Staff	X	X	
Warren Zepp	Plant Operations & Control	X	X	
Jim Muri	Plant Operations & Control	X	X	
Marcis Kempe	Operations Support/SCADA		X	X
Mark Johnson	Operations Support/SCADA	X	X	
Andrew Hildick Smith	Operations Support/SCADA	X	X	X
Valarie Moran	Operations Support/SCADA	X	X	
Dave Coppes	Operations Support/SCADA		X	
Greg DeFronzo	MIS		X	X
Bill Fahey	MIS		X	X
Mike Farmer	MIS		X	X
Mary Lydon	MIS		X	X
Pak Yen Lim	GIS	X	X	
Daniel Nvule	Planning		X	X
John Gregoire	Planning	X	X	
Stephen Estes-Smargiassi	Planning		X	X
Jae Kim	Capital Engineering	X	X	
Bill Brutsch	Operations Management		X	
Maria Orfeo	Executive Office	X	X	

SCADA SYSTEM

Massachusetts Water Resources Authority currently has a Supervisory Control and Data Acquisition (SCADA) system in place for control and data acquisition. The Human Machine Interface (HMI) is manufactured by Intellution (FIX32) and the operating system is Windows NT. The HMI system has 12,000 Tags (inputs and outputs) with 24 Intellution nodes. A small amount of real time and historical data is available to managers in a PI database. The PI system is being expanded for managers to have more access to 2000 data points. The data available to the managers is also collected in 15-minute intervals for long-term storage and stored on the Management Information Systems (MIS) Oracle database. The controllers in the water plant for collecting field inputs/outputs and for process control are Allen Bradley Programmable Logic Controllers (PLC) communicating to the to HMI. There are over 27 current PLCs with plans for another 111 PLCs in the future. Telog data recorders are utilized for collecting data at the metering (billing) sites. The Telog data recorders are now communicating to the same HMI Intellution system and the Telog data recorders are deployed at 225 sites. The MWRA system is very large and covers a large geographical area. The HMI SCADA system collects the data from a very sophisticated communication system that includes leased line, radio, and microwave. The SCADA system is on its own Local Area Network/Wide Area Network (LAN/WAN) with a connection through a gateway to the MIS WAN network.

The primary use of the SCADA system is for operations to monitor and control plant functions. The use of SCADA real time and historical data by managers and other non-operations personnel is more advanced and utilized than most utilities. MWRA has a complete SCADA vision and has written standards for upgrades and future plants. MWRA has balanced the need for new technology with budget dollars by interfacing existing Telog equipment to the new SCADA HMI without a wholesale replacement. MWRA has a training program in place for the SCADA system that allows the engineers and technicians to grow as the SCADA system is growing. The waterworks communication system is very diverse allowing radios to be deployed to eliminate the recurring lease cost where possible and utilizing the state microwave communication system backbone with leased lines where required. The hybrid communication system combines the need for high speed and reliability while minimizing the yearly O&M cost. The maintenance personnel are trained to maintain their communication system without maintenance contracts to specialized contractors. The MWRA SCADA system is the most utilized system on a company wide basis from personnel other than operations that has been witnessed on this AwwaRF project or other experiences.

OBSERVATIONS

MWRA has made great progress in the processing, use and dissemination of information in the past few years. MWRA is unique in its commitment and standardization of information reporting on a monthly and quarterly basis. Data are aggregated in a very structured and standardized manner so that these reports become of unique value in content and timing. Our perception of the capture, management and aggregation of information is presented in Figure C.6.

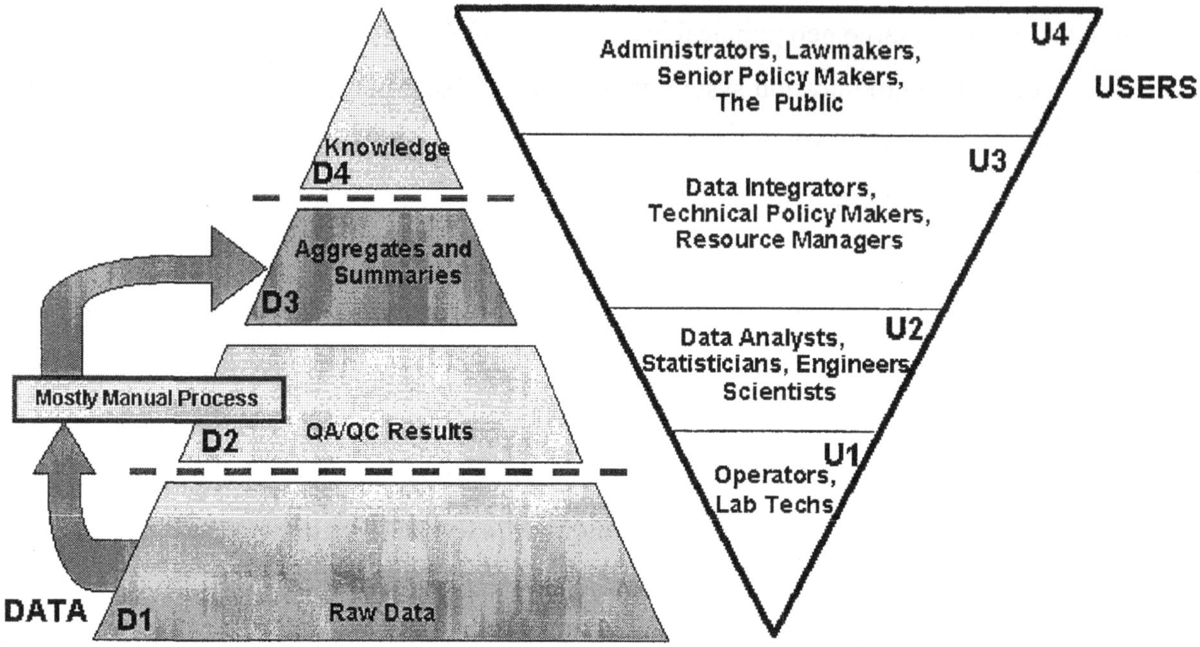

Figure C.6
Conceptualization of information capture processing and aggregation within MWRA

- One of the most insightful and productive features observed during the focus group meetings was MWRA's formalization of what we have been calling the Information Bureau. MWRA has empowered a number of committed people to work with all other organizations and to have a strong understanding of the available information and the key contacts for all the required information. This de facto Information Bureau made up of the Southborough data management staff, Jim Muri, Betsy Reilley-Mathews, Mike Farmer and Steve Rhode, are a tremendous resource that makes the formal creation of weekly and monthly (yellow book) reports possible. This represents the formalization of a concept that has been recommended at the other utilities in the case studies. We observed a similar commitment emerging at East Bay Municipal Utility District (in part based on observations made of the process used at MWRA). The Information Bureau personnel have a good integrated knowledge of the available data but this is not formalized in a directory of information resources. This deficiency results in utility personnel regularly communicating with the Information Bureau personnel when they need to locate data and information.

- While MWRA excelled at information aggregation, capture and organization of raw data was not as consistent. Different organizations have different methods and priorities for the processing and management of the raw data. Some of the groups have very sophisticated systems for capturing and managing raw data while other groups have not developed cohesive electronic capture and management of data. In some cases, data are retrieved on a schedule by telephoning client utilities. Flow data of tributaries is obtained from USGS (both on-line and via e-mail) and from the Army Corp of Engineers. Reservoir yields, land use data and watershed protection data are obtained from Metropolitan District Commission (MDC) either in hard copy or in spreadsheets. Systematic methods to store and share these data throughout MWRA are currently not available. This lack of standardization results in many different users within MWRA accessing the same data from outside sources. This works well for the users but in many cases requires redundant processing of the data. Different aggregation methods could result in different summaries for the same data by different departments. If these users were identified and access coordinated, all users could save time and ensure that the same results are being used across the entire utility.
- The aggregates and summarization in some cases drive the data collection at the D1 (raw data) level. The quality assurance and control of the data are inconsistent in an electronic form but the data required for the aggregate reporting are well done because of the high level of visibility of the aggregate monthly and quarterly reports. MWRA is currently reliant to a large degree on "cut and paste" of data from reports and spreadsheets maintained in individual divisions for the creation of the weekly, monthly and compliance reports. This strategy has a weakness related to transcription error that is inherent in significant "cutting and pasting". This is a recognized weakness that is slowly being addressed. Even with these limitations, a culture of weekly and monthly reporting has been developed and has resulted in high quality data aggregates at the D3 level (Figure C.6) but in many cases the electronic

QA/QC of the data and the intermediate databases that would normally be associated with D2 are absent.

- MWRA has been very successful at getting SCADA data available to a wide array of user through the use of the PI database. PI is an effective data historian that only creates records when changes of a certain pre-established magnitude thresholds are exceeded. It allows MWRA to store extensive amounts of data with very efficient storage and rapid access to a wide array of data. This system is used very effectively to support the operations and planning processes within the utility. Some concerns were expressed regarding the quality of the SCADA data and the accessibility of tools for analyzing and visualizing these data.
- MWRA is using the available SCADA data in PI and the data aggregates (weekly and monthly reports) for process benchmarking with other utilities. These data are being used to promote continuous improvement by comparing processes and outcomes with other drinking water utilities of similar size, sources and treatment.
- MWRA does not currently have a unified electronic data processing environment. While there is the unifying aggregation of data into information that makes up the weekly, monthly and compliance reports there is no plan or strategy for capturing, processing and storing this information in a organized cohesive form that will result in the information being available in a common database for both primary and secondary use. While knowledge is being developed within the utility this knowledge is not being developed in a format that can be accessed and searched electronically. This is greatly complicated by a lack of standards including naming conventions for:

  - Attribute names (column headings in databases)
  - Entities (table names)
  - Locations (e.g., for sampling)
  - Analyte names

- There are standard names used by segments of the organization but other segments of the organization may have different standard names for the same locations, attributes and analytes.
- MWRA again is unique in its recognition of the importance and in the application of records management to many of its critical reports and communications. This includes the storage and maintenance of both electronic and paper documents. As we have seen at most utilities, the capturing and organization of "as built" diagrams is a major issue within MWRA. Once the "as-builts" are entered there are not standard operating procedures in place for getting these modifications entered into the GIS. This failure to capture the "as-builts" in a timely manner was cited as a serious problem that has significant costs associated with it. For example, this may happen when crews are sent out to dig and they do not find the resources in the expected place. Tracking and maintenance of communications and permits were also cited as requirements that are being addressed by on going efforts in enhancing records management within MWRA.
- As part of our evaluation we explored the critical relationship between the MIS group and the departments that they support. This critical relationship is often the pivotal issue in the ability of a utility to develop an integrated information system. In general, the other groups within MWRA view the MIS group in a positive light. Most of the groups interviewed rely heavily on MIS for infrastructure, software support, communications, backups and security. The perception of those working with the MIS group is that when they turn their attention to an issue it gets resolved effectively. There are times when the delay between a request and action are too long. This was most strongly expressed relative to purchasing and to addressing hardware and software issues.
- There was also a perception that there is no clear plan for the MIS group. This was definitely a misconception from the perspective of the project team, which found a strong strategic plan with a very clear vision. It appears that while the MIS group has this clear vision, it has not been communicated effectively to many groups within the utility. Many of the concerns expressed during this study would be

strongly addressed if the vision and strategic plan of Information Systems (IS) were clearly communicated throughout MWRA.

- GIS support at MWRA is outstanding. The GIS support group is not under the MIS group but rather it is independent and works within the Operations Division focusing its attention on Planning. Like other groups within MWRA, the GIS group depends on MIS for many fundamental services. While there is a bit of tension related to the special needs of GIS (like storage, bandwidth for applications and the need for high-end computing requirements), the overall perception of MIS is positive. The GIS group has developed methodologies for processing data and maintaining the infrastructure coverages that appear to be unique among our case studies. The GIS group appears to be service oriented and has made significant strides relative to other utilities that we have interviewed.
- Many participants expressed a need for additional training for existing tasks using existing software tools. The major skills that need enhancing within the MWRA user community are related to data extraction, database design and programming (both database and programming languages and data analysis). There are some additional tools that are needed relative to these requirements as well, in particular data visualizations tools and a general ad hoc report writer.
- There were concerns expressed by both users and MIS about the heterogeneous information systems within MWRA. These heterogeneous systems include a mixture of DOS, UNIX and VMS based systems. These disparate systems often complicate the access of information and certainly complicate the networking and maintenance of the utility wide system. MWRA has planned a consolidation of operating systems by the end of 2002.
- MWRA is currently not submitting compliance data electronically to the regulatory agencies. This would be desirable and could be done based on the monthly reports and the compliance reports that are currently being done but the data submission would still not be automated. Similarly there is little electronic reporting being done to clients. Some of the weekly, monthly and quarterly reports are being transmitted electronically (via e-mail) but in all cases they are being aggregated and generated through manual or "cut and paste" processes.

- Maintenance management information was also cited as an important requirement by a number of different departments. Currently maintenance management information is only available via paper, e-mail and personal communications. However, a maintenance management system will resolve many of these issues as evaluation and implementation of Records Management has been implemented.

BARRIERS

In this section we summarize barriers participants identified impacting the access and use of information within the utility. These barriers are often symptomatic of larger issues. In some cases, the barriers may be easily overcome through some simple improvements or communications. In other instances, the barriers are substantial and will require changes in infrastructure, management, culture or operating procedures. The solutions to these barriers are addressed in the recommendation section. Many of these barriers have been discussed in the context of observations above.

- A major barrier to information access within MWRA is the lack of awareness of what information is available. The requirements analysis identified a significant amount of information available within MWRA that many employees are interested in accessing but do not know that the data are available or who to ask for the data.
- It is also difficult to obtain data access in many cases as well. Much of the necessary information is still maintained in paper format within filing systems accessible by a limited number of users. Other data is available on-line but in a wide variety of formats, in local drives or on public drives that few users know about.
- Many databases (i.e., Microsoft Access) have been created without consideration of standard naming conventions or the development of user interfaces. In some instances, data have been available in systems that users know about but could not access. An example of this is the Laboratory Information Management Systems (LIMS) data; however, this has been resolved recently by the creation of WQ2 and WQ3.

- In some instances, appropriate software for accessing, analyzing and interpreting data has not been available. Some of these restrictions, like access to the Internet, are organizational policy while other issues like software purchase has been a problem because of organizational bottlenecks. MWRA is very dependent on data from other agencies and some restrictions on accessing the Internet prevent staff from accessing required data. In some cases, data that were created in the past were developed using software tools that are no longer supported or that have been removed because of organizational standardization (i.e., lotus 1-2-3 files and data). The lack of training in the available tools has also been a barrier for the storage, access and analysis of data.
- Due to the distributed nature of MWRA, there are some difficulties accessing data in remote areas because of bandwidth and access issues. These are being resolved by MIS, but few users know the plans and schedules for improving data communications and access. To some degree, the MIS department is a victim of its own success. The users within MWRA have grown to depend on MIS and to expect good services. As the requirements have increased, the MIS group has not grown to meet the new demand. From the perspective of the MIS group it has been difficult to justify the burden to ratepayers to meet the needs of the MWRA staff. This has resulted in frustration to both the MIS group as well as to the user community within MWRA.

RECOMMENDATIONS

This section includes specific recommendations that the project team thinks would be helpful for MWRA to improve the capture, use and communication of information. MWRA has made great progress in the past few years in their information processing and usage. Some of the recommendations below were underway when they were originally presented to MWRA. Others have begun since the focus group meetings. Many of the recommendations made in this section were communicated at the time of the focus group meetings.

- MWRA is definitely a candidate for the use of knowledge management techniques including data mining and data warehousing. However, these tools will require a significant effort in the development of integrated information systems before they can be implemented. To some degree the data warehousing tools that are available could help in the integration of the disparate datasets and in information aggregation across different database designs and naming standards. These barriers make the implementation and maintenance of a data warehouse more difficult.
- MWRA has implemented a complete SCADA system with standards and schedules. The system is working efficiently and has no obvious need for major improvements beyond the planned implementation of upgrades. Since MWRA is planning to use the new object based release of Intellution (iFix), training in the Visual Basic Programming tool (VBA) is suggested for the engineers and technicians involved with this upgrade or responsible for on-going maintenance. The Intellution iFix product utilizes VBA scripts for the dynamic objects in the graphics such as Auto/Manual, Start/Stop and "Pop-up" screens. This is a more powerful tool than is currently being used; however, the use of VBA is new and a skill not previously required of SCADA personnel on a daily basis. In addition to the training, VBA script standards will need to be generated before deployed or implemented. The current Intellution Fix32 graphic screens can be imported to iFix and most objects will import without modification. However, the control scripts for the buttons will not import and the Intellution software will give a message screen of the specifics that need correction. Clearly articulating a standard will accelerate the installation

and ensure that whatever new programming is required will be maintained to keep present standards in place.

- The Information Bureau personnel have established contacts in most departments and now act as resources to many technical users for the access and analysis of data and information. This has resulted in wider use of data in MWRA than in many of the other utilities that we investigated. This role should be further formalized and the individuals that have emerged as the Information Bureau should be given additional tools, resources and training to allow them to continue improving the access and use of data.
- Largely due to needs of the Information Bureau, increasing bandwidth to the Southboro network should be made a priority.
- In general, the information aggregation that occurs is completed manually with a single goal in mind: that the required aggregates and visualization (tables, graphs and figures) be ready for the weekly and monthly reports. These aggregates are generated manually through extensive amounts of "cutting and pasting". Currently, many of the information aggregates that are produced are available only in report format. Many of these results could and should be generated electronically and then should be housed in aggregate formats in formal information systems for long-range access and integration into other information products. These aggregates can then form the basis of knowledge that is at the top of the data triangle in Figure C.6.
- The Information Bureau personnel need to have greater access and priority within the IS group to focus on their requirements. The current management alignment of the Information Bureau is appropriate. These resources need to be focused on the technical challenges of treating and delivering drinking water and not on information systems. This team should be closely aligned with the technical personnel that are analyzing and interpreting data, not with the information system personnel. However, they need to be supported extensively by the MIS staff in order to be successful. We would recommend the stationing of an IS person at the western operations center to focus on supporting this group. Appropriate members of the IS group should also spend two cycles of preparing the weekly, monthly and quarterly

reports to fully understand the requirements and the barriers to efficient information aggregation that the technical staff faces.

- A requirements analysis should be run with the Information Bureau group to determine the tool set that would best support their activities. The required tools should be purchased or installed for them and intensive training in database design, data extraction should be given to this group.
- The Information Bureau should be charged with the development of a directory of information resources and key contact people for accessing and applying data. These tools should be made available to the entire utility.
- The Information Bureau personnel should carefully document both the processing that is currently done and the quality assurance and control that is performed. The aggregate information required for the weekly, monthly and quarterly reports should be documented and the information flow to get the information should be documented. These evaluations should include any compliance reporting currently done or anticipated in the future. The Information Bureau personnel should work closely with the MIS group to evaluate the options for automating these processes and storing the results in aggregate databases. The results of the Information Bureau processing should be saved in relational databases that could form the foundation of a utility-wide water quality database. The data in this database (D2 level, Figure C.6) could then be the foundation of the weekly, monthly, quarterly and compliance reports. For example, the plant quality data is manually checked for validation by observations from Process Control staff and plant operators before use in the monthly reports. The data could be QC'd within the SCADA software or at a historian level. This would ensure accurate data in the monthly and executive reports. Currently, the weekly and monthly reports are generated by "cut and paste" methodology from different systems including SCADA. The data aggregated from SCADA could be directly exported to the common database for the monthly reports to eliminate manual entry or keypunch mistakes. Common tag names and naming conventions are utilized within the SCADA system. This technique could be used with the other systems including GIS and LIMS to have correlation between similar data.

- The data included in the reports could form the structure and content of the information aggregates. Again, these results including summary data, charts, tables and graphs could be the contents at the D3 level (Figure C.6). This information would be an ideal foundation for a web-based interface for the entire utility. Some subset of this data would also be appropriate for distribution to client utilities. This information base could also facilitate the automation of electronic weekly, monthly and quarterly reports. This may also allow the automation of compliance reporting. At a minimum, the weekly, monthly, quarterly and compliance reports should be stored electronically and accessible via the Wide Area Network.
- If the aggregation and automation of reporting is attractive and these recommendations are initiated, MWRA should begin discussions with the appropriate regulatory agencies as soon as possible to encourage them to allow electronic reporting of data. This is important to ensure that the regulatory agency requirements are implemented both from a format and content perspective as well as a communications strategy perspective (format, content and approach for communicating the data from the utility to the regulatory agency).
- MWRA is ready to move to the next level of data management and usage, which will be a significant quantum leap. It will involve the development of an integrated information system that accepts data from a number of different departments into a cohesive information system design. In order to do this, a significant effort is required to understand all of the detailed information system requirements and data flows. While this effort is related to the requirements analysis discussed above for the Information Bureau it is not the same thing. This requirements analysis will identify the needs of each of the groups, which is a superset of the requirements of the Information Bureau. It will establish the detailed information needs at the D1 level (Figure C.6). If the requirements are ever addressed through the implementation of an electronic system it will certainly support the creation of reports and data aggregates but it will also supply the detailed data required within each department to perform daily tasks. This requirements analysis will also help identify secondary uses of the raw data and will identify the detailed data verification and validation needs related to data collection. There was no clear

indication that an effort of this kind has occurred across the utility. Some individual departments have done this kind of in depth study in preparation for the procurement or implementation of new information systems. This effort could lead to the implementation of a cohesive integrated information system. This will require a significant investment, which needs to be carefully considered from the perspective of return on investment. In addition, this will require some management decisions. The current MIS director is responsible for supplying an operational information system environment. This includes hardware, software and support. This is a large and complex responsibility. A major requirements analysis and consideration of integrated information systems is a large undertaking. It is certainly related to the MIS mission, but too much to ask the existing staff and leadership to undertake. Additional resources would be required to work either in coordination with or at the direction of the MIS director in order to accomplish this task. Successful implementation will require a strong senior manager with the authority to impose standards and standard operating procedures.

- Implementation of integrated information systems and possibly knowledge management tools is an expensive proposition and requires not only significant investments in both money and staff time, but also requires a change in organization culture to be effective. This cultural change is often more difficult than the investment and the challenge of gathering and organizing all of the required information. A commitment needs to be made to becoming an information-based organization. Standard operating procedures for data entry, processing and documentation must be adopted. Perceptions that data belong to one part of the organization to the exclusion of others is not acceptable and needs to be clearly and aggressively discouraged by senior management. System implementation needs to include extensive training in the specific application of the system especially for middle and upper managers.
- Regardless of whether or not the integrated information system is pursued, MWRA should immediately convene a standards working group to identify required standards and the ad hoc solutions that are currently being used to meet these needs. The group should then begin implementing standards. Wherever possible existing or

industry wide standards should be given serious consideration. This group should pay attention to the AwwaRF project (2764) that is currently being evaluated for funding since it will be working on the development of water quality database standards for the drinking water industry. Standard development should be accompanied by extensive input and review within the utility. Once standards are established they should be implemented. An implementation plan should include the schedule for implementation and impacts within the utility. Any introduction of a new standard is likely to have far reaching impacts within MWRA. These need to be carefully considered and resources need to be expended to support this implementation.

- The MIS department should institute a review of their internal processes for requesting support. Careful attention should be given to procurement of hardware and software and the subsequent installation of these components. The goal of this review should be to improve the efficiencies and cut down on the red tape. Some of the red tape may be appropriate and required but wherever possible it should be reduced to optimize the resources used within the MIS department and to reduce the frustration of the user community.
- The MIS department and the rest of the utility would benefit from an aggressive communication of the MIS strategic plan to the rest of the utility. This could be done a number of ways including a traveling presentation or an on-line summary of the strategic plan. Many of the concerns expressed by the user community are being addressed or will be addressed shortly. This focus and schedule is not widely known by those outside the MIS group.
- MWRA has recognized the importance of record management. This is a requirement that we have identified consistently at all utilities. MWRA should evaluate existing records management packages and implement an updated off the shelf system. This system should include a web-based interface to allow wider access to more utility personnel. The records management system should facilitate the use of scanned documents, designs communications, e-mails, presentations and the use of aging tools to ensure that once a document can and/or should be deleted it is purged, archived or destroyed.

GLOSSARY

Algorithm: A set of rules or procedures (e.g., a detailed order of actions to perform on data) that enable problem solving. In computer science, an algorithm (e.g., a computer program) usually refers to a method that solves a recurrent problem.

Business Intelligence: Business Intelligence is a set of concepts, methods and processes to improve business decisions using information from multiple sources and applying experience and assumptions to develop an accurate understanding of business dynamics. It is the gathering, management, and analysis of data to produce information that is distributed to people throughout the organization to improve strategic and tactical decisions.

Business Intelligence (BI) Tools: Applications that enable business users to see, use and analyze large amounts of complex data. OLAP, Query tool and reporting, and data mining are the most commonly used BI tools.

Business Intelligence (BI) Systems: Systems especially designed to "get information out" and are built to allow exploration, analysis, and presentation of information. BI systems involve integrating tabular data with text documents and graphics and allow desktop users to access current data and information along with historical information and retrospectives regarding the success or failure of past decisions. ***See Operational Systems.***

Central Data Warehouse: Central data warehouses use data directly from data archive systems and other sources and can transfer data to departmental data warehouses (e.g., data marts). ***See Data Warehouse and Data Marts.***

Computer Output to Laser Disk (COLD): An application that provides fast, random access to an enormous amount of data stored on RAID cache, optical disk or other large storage device.

Corporate Document Portal: An access point for the capture, manipulation, retrieval, distribution and interaction of corporate documents.

Customer Relationship Management (CRM): Process and applications by which a company maximizes customer information in an effort to increase loyalty and retain a customer's business over its lifetime.

Data: Single piece of information such as names, dates or values made from observations. Data are raw values and do not provide for a means of interpretation.

Database Management System (DBMS): Software that is used to store, access, and manage data.

Data Mart (Departmental Data Warehouse): A database that has the same characteristics as a data warehouse but is usually smaller and is focused on the data for one division or department within an enterprise. Independent data marts use data directly from data archive systems and other data sources. Data marts contain much smaller volumes of data than a central data warehouse and have fewer interrelationships. ***See Data Warehouse and Data Mart.***

Data Mining: A process by which data trends (hidden patterns and relationships in the data) can be identified and evaluated. This process turns data into information through the use of visual representations, statistical analysis, modeling, and assessment of the results. Data mining uses the data to answer questions, which frequently lead to more specific questions through an iterative process.

Data Mining Tool: A computer application (software) that automatically searches for patterns and correlations in data.

Data Warehouse: Process and tools used to consolidate data from various sources into an environment optimized for analytical and information processing. Data warehousing involves the entire information process as it accesses, manages, organizes and utilizes data through tools including OLAP, data mining and query and reporting tools.

Decision Support System (DSS): A computer system designed to assist an organization in making decisions.

Document: Information that is essential to a company and may reside in numerous locations, including on-site and off-site locations, document archives, etc. ***See Document and Record Management.***

Document Imaging: Document imaging converts paper documents into electronic formats by organizing, scanning, indexing and storing this information in an electronic document management system.

Document File Management: Document file management involves the creation, capture, indexing and retrieval of many file types including word processing documents, spreadsheets, TIFFs, faxes, and PDFs.

Document and Record Management: An application that manages, stores and distributes documents and records throughout their lifecycle. It also adds version control, shared workflow and structured communications. In general, document management focus is short-term whereas records management focus is more long-term. ***See Document and Records.***

Electronic Document Management: An electronic integrated system used to create, store and manage documents (e.g., text, video, databases, etc.) which utilizes many different technologies (e.g., scanners, printers, servers, relational databases, etc.).

Enterprise Information Portal (EIP): Enterprise Information Portals (EIPs) provide total access (essentially a single gateway) to multiple forms of critical data as a starting point for a user (e.g., customer, employee, etc.). This type of portal acts as a transition, pulling from mainframe systems and presenting the information within a browser window and some even support e-mail and scheduling functions.

Enterprise Resource Planning (ERP): Process and applications (software) that enable organizations to better manage resources and allow the integration of daily operational functions (e.g., enabling human resources, payroll, and financial systems to relate to one another and share data).

Geo-Spatial: Spatially oriented information that can be displayed geographically.

Geographic Information System (GIS): A computer system capable of assembling, storing, manipulating and displaying geographically referenced information (e.g., data identified according to their locations) that can be used in environmental investigations, water resource management, watershed monitoring and development planning.

Hybrid OLAP (HOLAP): A combined use of Relational OLAP (ROLAP) and Multidimensional OLAP (MOLAP).

Information: Organized raw data that has meaning and delivers a message in a structured format. Both soft networks (e.g., memo) and hard networks (e-mail, files, and wires) deliver information. Information generally has an influence on the person to which it is transferred.

Informational Metadata: Informational metadata describes the content or purpose of a table or the name of the responsible party for the information in the table. ***See Metadata.***

Knowledge: Knowledge takes raw materials (data and/or information) to the next level and can be viewed as a finished product. Knowledge can be contextual information contained within documents and archives, databases, policies and procedures, personal experiences and expert insight.

Knowledge Management: Knowledge Management can be viewed as a discipline that promotes an integrated approach of combining technology and strategies and provides for adaptation and survival in a rapidly changing environmental workplace. The goal of Knowledge Management is to identify, capture, manage, analyze and extract knowledge from various sources within an organization and place it into a structured, organized format (which may need to be refined over time) and can be shared.

Laboratory Information Management System (LIMS): An application (software) that is utilized by water utility laboratories for storing and reporting water quality data. It also serves as a repository of water quality information, allowing users to observe trends in data. Data is stored and retrieved in a way that allows laboratory staff to track the progress of a given batch of samples and data users to receive answers to water quality inquiries.

Management Information System (MIS): An information system at the management level of an organization that combines data and analytical models to support decision-making.

Metadata: An index or template that describes electronic information on a server using data fields. One example of a metadata application is a library card catalog.

Multidimensional Analysis: A process of analysis that involves organizing and summarizing data in a multiple number of dimensions.

Multidimensional Databases: Multidimensional databases group fields into dimensions, and then dimensions into multidimensional arrays. ***See OLAP.***

Multidimensional Database Management System (MDBMS): A database management system (DBMS) that organizes data multidimensionally.

On-Line Analytical Processing (OLAP): An analytical information processing technology able to examine data from multiple dimensions "on-line" and interactively. In general, OLAP can be viewed as a hardware/software configuration that works as an observation and query tool used to uncover new relationships from the existing data and as a support tool for analysis and confirmation at the review level. Although OLAP tools present data to the users in a multidimensional fashion, the databases that support this technology can be relational, multidimensional, and vertical. ***See Relational, Multidimensional, and Vertical Databases.***

Operational Management Systems (OMS): Applications that manage operational systems. ***See Operational Systems.***

Operational Systems: Hardware/software configurations designed to "get data in" and are utilized to automate routine tasks and are characterized by large volumes of small transactions (transactional data). ***See Business Intelligence.***

Physical Metadata: Physical metadata defines the way an application stores or transforms data, such as the number or rows and columns within a table or any type of transformations applied to those rows or columns. ***See Metadata.***

Portals: Web applications that allow electronic document retrieval for customer self-service, business documents (e.g., contracts, policies, forms, bills, and invoices), public document access and web posting. In addition, they also include e-mail and content capture, warehouse and folder management, as well as correspondence and case management.

Query and Reporting Tools: Computer applications (software) that allow the user to ask questions about patterns or details in the data.

Redundant Array of Inexpensive Disks (RAID): Hardware (a bank of hard disks) used to increase system reliability storage. Data is organized on the disks such that the failure of any single one does not cause a total loss of data. This enables the system to continue in use while the faulty disk is replaced.

Records: Records can be any information that is essential to a company and may reside in numerous locations, including on-site and off-site locations, document archives, etc. Records need to be stored and sometimes retrieved.

Regression: A statistical technique used to investigate potential relationships between two variables.

Relational databases: Databases that organize data horizontally in rows called records. Each record contains related information that is organized into fields. ***See OLAP.***

Relational Database Management System (RDBMS): A database management system (DBMS) based on the relational theory. Information is stored in separate tables, which can then be related. Most modern DBMS such as Oracle, Sybase and Microsoft SQL Server are included in this category. RDBMS are also used for multidimensional analysis.

Scale, Scalable, Scalability: Having to do with the ability of a computer or a database to operate efficiently with larger quantities of data.

Statistics: The branch of applied mathematics that identifies relationships and searches for ways to understand these relationships. When planning a scientific experiment, statistics need to be considered in all aspects including the collection of scientific data, subsequent data management, drawing inferences from data and the presentation of results. For example, descriptive statistics are a means of summarizing observational data through the calculation of a mean, mode, average, standard deviation, variance, etc.

Structured Query Language (SQL): The standard language for accessing relational databases. Structured Query Language syntax is a query and reporting tool that allows for the creation of queries enabling the user to search for patterns or trends in data.

Supervisory Control and Data Acquisition Systems (SCADA): Systems used for the monitoring and control of operating networks in water systems, as well as management of wastewater processing facilities.

Vertical Databases: Vertical databases are based on data that is stored vertically in columns. ***See OLAP.***

Workflow: The flow of information in a business process.

REFERENCES

American Productivity & Quality Center (APQC). 1996. Knowledge Management.

American Productivity & Quality Center (APQC). 2000. Successfully Implementing Knowledge Management.

Barnicle, A. 1997. Implementing Monitoring, Control and Management Information Systems in a Major Water Utility. Water Supply, 15:2, pp. 75-99.

Burchfield, V. and M. Harston. 1999. Columbus Water Works Integrates Applications for Improved Reporting and Management. AWWA Information Management and Technology Conference (IMTECH). April 18-21, 1999. New Orleans, Louisiana. American Water Works Association.

Butler, S. 1999. Knowledge Management Directions: From BI to Knowledge Management. Data Management (DM) Review.

Casey, R.J. and Grinney, S.A. 1993. An Operational Information Database Based on the Automated Processing of Time Series Available From Telemetry Control Systems. Water Science and Technology, 28:11/12, pp. 111-121.

Cole, L.A., C.J. Stansberry Jr., K.D. Le, and H. Ma. 1996. Client-Server Technology Meets Operational-Planning Challenges. IEEE Computer Applications in Power. p. 45-49.

Cromwell, J., D. Owen, J. Dyksen, and E. Means. 1997. Managerial Assessment of Water Quality and System Reliability. Published by the AWWA Research Foundation and American Water Works Association. ISBN 0-89867-905-2.

Davis, J. and Joyner, E. 1999. Successful Customer Relationship Management: Why ERP, Data Warehousing, Decision Support, and Metadata Matter. A SAS Institute White Paper.

Fayyad, U. and Uthurusamy, R. 1996. Data Mining and Knowledge Discovery in Databases. Communications of the ACM, 39:11, pp. 24-26.

Firestone, J. M. 1997. Evaluating OLAP Alternatives. A White Paper from Executive Information Systems, Inc.

Firestone, J. M. 1998a. Architectural Evolution in Data Warehousing and Distributed Knowledge Management Architecture. A White Paper from Executive Information Systems, Inc.

Firestone, J.M. 1998b. META Group and the Future of Knowledge Management. A White Paper from Executive Information Systems, Inc.

Firestone, J. M. 1999a. Enterprise Information Portals and Enterprise Knowledge Portals. A White Paper from Executive Information Systems, Inc.

Firestone, J. M. 1999b. Defining the Enterprise Information Portal. A White Paper from Executive Information Systems, Inc.

Haake, K. 1999. Comparison of three GIS Applications to Serve Information on a Water Utility Web Site. AWWA Information Management and Technology Conference (IMTECH). April 18-21, 1999. New Orleans, Louisiana. American Water Works Association.

Hertz, C., Bardsley, A., Kurisko, P. and Senft, M. 1998. System Helps Agency Manage Laboratory Information. Waterworld, October.

INFORMIX. 1998. An Integrated Approach to Enterprise Resource Planning and Data Warehouse Systems. An INFORMIX White Paper.

Morrow, P. and J. Wrigley. 1999. Pinellas County Utilities: Leveraging Technology to Assist with Infrastructure management. AWWA Information Management and Technology Conference (IMTECH). April 18-21, 1999. New Orleans, Louisiana. American Water Works Association.

Paraira, M. and Tejero, J. 1999. How Water and Environmental Laboratories Benefit from LIMS Software Installation. International Environmental Technology, July/August.

PILOT SOFTWARE. 1999a. An Introduction to OLAP Multidimensional Terminology and Technology. A PILOT Software White Paper.

PILOT SOFTWARE. 1999b. Customer Intelligence: Business Intelligence for CRM Solutions. A Pilot Software White Paper.

SAS INSTITUTE. 1996a. OLAP tools and Techniques within the SAS System. Technology to Match the Way People Think. A SAS Institute White Paper.

SAS INSTITUTE. 1996b. A Formula for OLAP Success. Using On-Line Analytical Processing to Transform Data Into Knowledge. A SAS Institute White Paper.

SAS INSTITUTE. 1997. Data Warehouse Administration. Produced in Association with the Data Warehouse Network. A SAS Institute White Paper.

SAS INSTITUTE. 1998a. Data Mining and the Case for Sampling. Solving Business Problems using SAS Enterprise Miner Software. A SAS Institute Best Practices Paper.

SAS INSTITUTE. 1998b. SAS Rapid Warehousing Methodology. A SAS Institute White Paper.

SAS INSTITUTE. 1999a. Finding the Solution to Data Mining. A Map of the Features and Components of SAS Enterprise Miner Software Version 3. A SAS Institute White Paper.

SAS INSTITUTE. 1999b. Implementing a CRM-Based Campaign Management Strategy. A SAS Institute White Paper.

SAS INSTITUTE. 1999c. The Quality Data Warehouse. Serving the Analytical Needs of the Manufacturing. A SAS Institute White Paper.

SAS INSTITUTE. 1999d. Intelligent Warehousing for ERP Systems: Leveraging SAP AG's R/3. A SAS Institute White Paper.

SAS INSTITUTE. 1999e. SAS joins the Institute for Knowledge Management. Press Release Archive, 18 August 1999.

Tekeitz, F. and McCarthy, R.L. 1999. Analytical Customer Relationship Management. A White Paper from Sybase, Inc.

TOWER TECHNOLOGY. 1998. Computer Output to Laser Disk: What is mean by the term "Output"? And does it Matter? A Tower Technology White Paper.

TREEV. 1998. TREEV: The New Paradigm in Document Management. A TREEV White Paper.

Thurman, S. 1999. Knowledge Management Means Career Opportunity. Intercom. 46:7 p.19-21.

U.S. Environmental Protection Agency (U.S. EPA). 1998 National Primary Drinking Water Regulations: Interim Enhanced Surface Water Treatment: Final Rule. Federal Register: 63(241):69478-69521.

Ward, G.D. 1996. International Report: Network Management Systems. Water Supply, 14:3/4, pp. 121-155.

Internet Sites

Baltimore Gas and Electric at: http://www.imagingmagazine.com/db_area/archs/1997/January/4_1002.htm.

Essex and Suffolk Water case study at: http://www.towertech.com/government/index.cfm?catid=00019DA4-D608-1A54-89A39EEF40F4DCF2&language=english .

Furnas Electric case study at http://www.infograph.com/doc/stories/furnas.htm.

General Electric case study at: http://www.towertech.com/government/index.cfm?catid=00033138-D63B-1A54-89A39EEF40F4DCF2&language=english.

Georgia Power Company case study at: http://www.towertech.com/government/index.cfm?catid=00033138-D63B-1A54-89A39EEF40F4DCF2&language=english

GPU Energy case study at: http://www.coherentnetworks.com/index-electrical.html

Gulf Power Company case study at http://www.infograph.com/doc/stories/gulf.htm.

Integral Energy case study at: http://www.towertechnology.com/solutions/apps_IE.htm.

Niagara Mohawk Power Corporation case study at http://www.coherentnetworks.com/index-electrical.html.

Northeastern States Power Company http://www.coherentnetworks.com/index-electrical.html.

Omaha Public Power District case study at:
http://www.onbase.com/solutions/casestudies/oppd.pdf.

Ontario Hydro case study at:
http://www.imagingmagazine.com/db_area/archs/1996/January/4_1256.htm.

OTAY case study at: http://www.infograph.com/doc/stories/otay.htm.
PacifiCorp case study at: http://www.coherentnetworks.com/index-electrical.html

Pennsylvania Power and Light Company case study at: http://www.coherentnetworks.com/index-electrical.html.

PSE&G case study at:
http://www.imagingmagazine.com/db_area/archs/1996/January/4_1256.htm.

South East Water Limited case study at:
http://www.imagimgmagazine.com/db_area/archs/1996/January/4_1256.htm.

Sydney Water case study at:
http://www.towertech.com/government/index.cfm?catid=000C7F81-D649-1A54-89A39EEF40F4DCF2&language=english

Wessex Water Services Ltd. Case study at: http://www.lucent.com/press/0200/000222.coa.html

Yorkshire Water case study at:
http://www.towertech.com/government/index.cfm?catid=00058CA2-D6AB-1A54-89A39EEF40F4DCF2&language=english

ABBREVIATIONS

AWWA	American Water Works Association
AwwaRF	Awwa Research Foundation
BI	Business Intelligence
COLD	Computer Output to Laser Disk
CRM	Customer Relationship Management
DBMS	Database Management System
DMWW	Des Moines Water Works
DSS	Decision Support System
EBMUD	East Bay Municipal Utility District
EDI	Electronic Data Interchange
EIP	Enterprise Information Portal
EMPACT	Environmental Monitoring for Public Access and Community Tracking
ERM	Enterprise Report Management
ERP	Enterprise Resource Planning
GIS	Geographic information systems
HOLAP	Hybrid On-Line Analytical Processing
IT	Information Technology
KM	Knowledge Management
KMS	Knowledge Management System
LIMS	Laboratory Information Management Systems
MDBMS	Multidimensional Database Management System
MIS	Management Information System
MOLAP	Multidimensional On-Line Analytical Processing
MSDS	Material Safety Data Sheets
OLAP	On Line Analytical Processing
OLTP	On-Line Transactional Processing
OMS	Operational Management Systems
OPPD	Omaha Public Power District

PAC	Project Advisory Committee
PDF	Portable Document Format
PSE&G	Public Service Electric and Gas
QA	Quality Assurance
RAID	Redundant Array of Inexpensive Disks
RDBMS	Relational Database Management System
RM	Records Management
ROLAP	Relational On-Line Analytical Processing
SCADA	Supervisory Control and Data Acquisition
SEMMA	Sample, Explore, Modify, Model, and Assess
SQL	Structured Query Language
SSD	Sewerage Service Diagram
TIFF	Tag Image File Format
TVA	Tennessee Valley Authority
U.S. EPA	United States Environmental Protection Agency
VOLAP	Vertical On-Line Analytical Processing
WAVES	Work Allocation

IWA Publishing's authorised EU representative for General Product Safety Regulations is Diane D'Arras, 15 rue Duret, 75116 Paris, France, e-mail: safety@iwap.co.uk.

Printed and bound by CPI Group (UK) Ltd, Croydon, CR0 4YY
17/09/2026
02218813-0006